A History of the Future

Prophets of Progress from H. G. Wells to Isaac Asimov

In this wide-ranging survey, Peter J. Bowler explores the phenomenon of futurology: predictions about the future development and impact of science and technology on society and culture in the twentieth century. Utilizing science fiction, popular science literature and the novels of the literary elite, Bowler highlights contested responses to the potential for revolutionary social change brought about by real and imagined scientific innovations. Charting the effect of social and military developments on attitudes toward innovation in Europe and America, Bowler shows how conflict between the enthusiasm of technocrats and the pessimism of their critics was presented to the public in books, magazines and exhibitions, and on the radio and television. A series of case studies reveals the impact of technologies such as radio, aviation, space exploration and genetics, exploring rivalries between innovators and the often unexpected outcome of their efforts to produce mechanisms and machines that could change the world.

PETER J. BOWLER is Emeritus Professor of the History of Science at Queen's University Belfast. He is a Fellow of the British Academy, a Member of the Royal Irish Academy, and a Fellow of the American Association for the Advancement of Science. He has published a number of books on the history of biology and several general surveys.

A History of the Future

Prophets of Progress from H. G. Wells to Isaac Asimov

Peter J. Bowler

Queen's University Belfast

CAMBRIDGE
UNIVERSITY PRESS

CAMBRIDGE
UNIVERSITY PRESS

University Printing House, Cambridge CB2 8BS, United Kingdom

One Liberty Plaza, 20th Floor, New York, NY 10006, USA

477 Williamstown Road, Port Melbourne, VIC 3207, Australia

314-321, 3rd Floor, Plot 3, Splendor Forum, Jasola District Centre, New Delhi - 110025, India

79 Anson Road, #06-04/06, Singapore 079906

Cambridge University Press is part of the University of Cambridge.

It furthers the University's mission by disseminating knowledge in the pursuit of education, learning and research at the highest international levels of excellence.

www.cambridge.org
Information on this title: www.cambridge.org/9781316602621
DOI: 10.1017/9781316563045

© Peter J. Bowler 2017

First published 2017

A catalogue record for this publication is available from the British Library

Library of Congress Cataloging in Publication data
Names: Bowler, Peter J., author.
Title: A history of the future : prophets of progress from H.G. Wells to Isaac Asimov / Peter J. Bowler, Queen's University Belfast.
Description: Cambridge ; New York : Cambridge University Press, 2017.
Includes bibliographical references and index.
Identifiers: LCCN 2017024135| ISBN 9781107148734 | ISBN 9781316602621 (pbk)
Subjects: LCSH: Forecasting. | Science fiction – History and criticism.
Classification: LCC CB158 .B69 2017 | DDC 303.49–dc23
LC record available at https://lccn.loc.gov/2017024135

ISBN 978-1-107-14873-4 Hardback
ISBN 978-1-316-60262-1 Paperback

Contents

Colour plates are to be found between pp. 150 and 151.

Plates

1. Poster advertising Alexander Korda's 1936 film of H. G. Wells's *Things to Come.* Alamy images.
2. Poster advertising the New York World's Fair of 1939. Alamy images.
3. Futuristic commuter train imagined in an advertisement for the Bohn Aluminum and Brass Corporation in the 1930s. From D. A. Hanks, *American Streamlined Design* (2005), p. 227. Reproduced by kind permission of the Syndics of Cambridge University Library (S950.a.200.1646).
4. Projected unmanned Mars probe. Frontispiece to Arthur C. Clarke's *The Exploration of Space* (1951). Reproduced by kind permission of the Syndics of Cambridge University Library (424.c.95.54).
5. Imaginary transatlantic rocket, from the front cover of Hugo Gernsback's magazine *Science and Mechanics*, November 1931. Getty Images. Note the similarity to the science-fiction rocket images of the period.
6. Imaginary airport on the roof of a skyscraper, from the front cover of *Meccano Magazine*, May 1932. Author's collection.
7. 1930s poster advertising the airline KLM. Alamy images. The Flying Dutchman of legend becomes a reality thanks to the progress of technology.
8. Poster advertising transatlantic flights by the airship *Hindenburg*, 1936. Alamy images.
9. Image representing Sir Ronald Ross (discoverer of the malaria parasite) inviting Europeans to colonize the tropics. From Arthur Mee, ed., *Harmsworth Popular Science* (1911), vol. 1, facing p. 233. Author's collection.

Figures

Preface

I have occasionally been asked why – having built my career researching the history of Darwinism – I have ended up writing a book about futurology. There are several mutually compatible explanations for the transition. One is that anyone interested in the origins of evolutionism knows that the theory was at first strongly linked to the idea of progress, and predicting the future progress of science is an obvious extension of that link. Another explanation emerges from the fact that I have steadily moved away from Darwinism by a series of steps, each of which seemed to make sense at the time. Studying the early evolution debates involved me with the whole question of science and religion, and I became increasingly frustrated by the fact that – for Britain at least – historians seemed to lose interest in that relationship at the point when the Victorian era came to an end. A study of the relationship between science and religion in early twentieth-century Britain led toward another area, that of popular science writing, again sadly neglected after the Victorian period. (I should add that neither of these lacunae are present for the situation in the United States.) This book emerges in part from my work on the British popular science literature of the early twentieth century, where I began to notice that predicting future developments seemed to become increasingly prevalent. Having always had an amateur interest in science fiction, it seemed natural for me to link the two areas of prediction and look for generalizations. The fact that both Wells and Asimov wrote popular science as well as science fiction illustrates the kind of synergy that interests me.

I have confined myself to the first two-thirds of the twentieth century partly because that seems to represent a reasonably coherent period for this topic (and also because I feel uncomfortable treating events I can remember as past history). As far as I know, there is no other book that seeks to survey predictions of the future on as broad a scale as that attempted here. We have many studies of science fiction and of the predictive novels written by literary figures such as Aldous Huxley. There are also many surveys of futurological speculation about particular areas of technical development and their implications. The use of such studies to throw light on cultural history is especially strong for America and a few other areas such as the Soviet Union. These studies make extensive use of popular science literature as well as science fiction and I have mined them shamelessly for information in what follows. One country for which the popular science literature does not seem to have been exploited is Britain, and here I have made use of my own detailed research. Since I read French, I have also made brief forays into the popular science literature in that language. I am acutely aware that even so, this is very much a survey based on the English-language literature, both primary and secondary. My only excuse is that at the age of 72 I don't feel up to trying to extend myself further.

I hope that this book will be of interest to a wide range of readers and I have been given to understand that it might prove useful as a text for a number of university-level courses. This has created something of a problem when it comes to providing documentation for the information and assertions contained in the text. Academic readers and students need references to the material from which evidence is derived, but ordinary readers may be put off by too obvious a display of erudition. The situation is compounded by the fact that in some areas I can appeal to reliable secondary sources to back up what I say, whereas for the British and French literature I need to provide detailed references to books, magazines and newspapers which have remained little researched up until now. Many of my references thus consist of an uncomfortable mix of secondary and primary sources, and of both academic and popular literature. To avoid littering the text with superscripts, I have in general provided all the documentation for each paragraph in a single note, and I apologize if these are sometimes rather complex. The bibliography is huge, reflecting both the breadth of the primary sources that bear on this topic and the richness of the secondary literature available in some areas. To keep the bibliography from getting really out of hand, references to newspaper and popular magazine articles are given only in the notes.

I'd like to end by recording the encouragement of several colleagues, including Amanda Rees, James Secord, Simon Schaffer and Charlotte Sleigh and the help of the editors at Cambridge University Press, Lucy Rhymer and Melissa Shivers.

1 Introduction

Progress or Threat?

After visiting the New York World's Fair in 1964, Isaac Asimov – already a well-known science fiction author – wrote a prediction for the *New York Times* of what he thought would be displayed at a similar exposition in fifty years' time.[1] When the new year of 2014 was ushered in, there was a flurry of media attention focused on the accuracy (or otherwise) of Asimov's vision. We seem to be fascinated by predictions of the future and also by a retrospective evaluation of earlier generations' efforts. It's almost as though we relish a demonstration of just how wrong earlier thinkers were about what we ourselves are actually experiencing. A popular book published in 2012 collected a host of articles and images from early volumes of *Popular Mechanics* under the title *The Wonderful Future that Never Was*.[2] Asimov actually did pretty well, although like many at the time he predicted a rosy future for nuclear power. More seriously, he combined optimistic forecasts about the future development of technology with a more sober assessment of what life might be like in an overcrowded and increasingly resource-starved world. Others at the time worried over the destruction of civilization or even of humanity itself by a nuclear holocaust. A companion volume to the *Popular Mechanics* survey highlighted predictions about new weapons that might be developed, showing that the optimistic and pessimistic visions of a technologically-rich future have always run side by side.[3]

Asimov was not just a famous writer of science fiction; he was also a prolific contributor to the genre of popular science writing. In this dual role, he paralleled the efforts of an even earlier pioneer, H. G. Wells, who predicted the future in novels such as *The War in the Air* and *The Shape of Things to Come*, but also made serious efforts to alert the public to the latest scientific developments (see colour plate 1). Wells and Asimov were both fascinated by science and technology, and were convinced that together these enterprises had the potential to change the world for the better. But they were also concerned that human nature and the dysfunctional state of modern society might lead to a disastrous misuse of technology. As science fiction writers, they imagined future worlds in order to tell stories about how human beings might cope with the challenges thrown up by new machines. And as popular science writers,

1

they were happy to engage in an imaginative form of an enterprise which would later be called futurology.

The term 'futurology' came into use in the 1950s to denote efforts to predict the future by extrapolating social and economic trends, increasingly via the use of computers to crunch the figures. The previous generation did the same thing on an individual basis. Writers like Asimov were trying to predict how society might evolve, but they were more interested in how social trends might be created or deflected by new technologies that they saw as plausible given the state of scientific knowledge at the time. This was a 'what if?' approach necessitated by the fact that one could only guess which of the potentially viable inventions would actually become successful. I am going to borrow the term 'futurology' to describe this project even though it is much more open-ended than the number-crunching of the global organizations that now worry about what might happen to us. Like science fiction, this kind of futurology makes an imaginative leap to pick out which possibilities will be explored, but doesn't set a human-interest story in the projected future. I want to argue that there is a close link between the two projects. *The Shape of Things to Come* was really this kind of futurology thinly disguised as fiction (although the subsequent movie version has a bit more action in it).

Popular interest in these predictions derives from our fascination with their accuracy or lack of it. We still worry that Aldous Huxley may have all too perceptively anticipated current trends in the application of biological technology in his *Brave New World*. The new technological gadgets foretold by Asimov and the *Popular Science* writers arouse admiration or contempt depending on whether or not we really are using them today. But for the historian of science or of popular culture, these anticipations of the future are a valuable resource for understanding the attitudes, beliefs and expectations of a generation that was getting used to the idea that the future would not merely repeat the past because science and technology were having irreversible effects on how we live. At the most basic level, they tell of the hopes for progress and the fears that it all might go horribly wrong. At a more detailed level, they tell us just what the technically savvy writers of the time thought might be the most fruitful lines of development and what they thought the effects would be for the ordinary person. Any effort to understand the interaction of science and technology with the emergence of modern social and cultural values must take these earlier prognostications into account.

Academic interpretations of early twentieth-century attitudes to science tend to focus on the pessimists. *The Shape of Things to Come* predicted a war that would almost destroy civilization, although at this point Wells still thought that a rationally planned state would emerge when the few remaining scientists took over and began to rebuild. The later 1930s saw a plethora of novels anticipating mass destruction through bombing, poison gas and even germ warfare. *Brave*

New World suggested that if the world ever did become governed by technicians, it would become a nightmare as they moulded humanity into a herd of
passive consumers. C. S. Lewis also turned to a cross between science fiction
and fantasy in his trilogy parodying Wells's vision of a scientifically planned
world. Many recent surveys of the period's efforts to foretell the future concentrate almost exclusively on the literary intellectuals' and novelists' efforts to
convince their readers that technology would destroy them one way or another.
The chapters on the early twentieth century in I. F. Clarke's *Pattern of
Expectation* paint a picture of unrelieved gloom derived almost exclusively
from the fiction of the time.[4] More recently, Richard Overy's *Morbid Age*
trawls though the writings of intellectuals and political thinkers to paint
a similar picture of a period when almost everyone thought that a catastrophe
was imminent.[5]

In 1940, George Orwell commented on the pessimistic attitude of literary
intellectuals such as Huxley, noting: 'All of them are temperamentally hostile
to the notion of "progress"; it is felt that progress not only doesn't happen, but
ought not to happen.'[6] Orwell soon became more pessimistic, but at this point
he thought the literary figures of the interwar years were out of touch with what
was really going on. He knew that the scientists, technical experts and
designers of the period were predicting the next steps forward in a much
more positive light. Huxley parodied not only Wells's plan for a technocratic
world state, but also the predictions of eminent scientists such as
J. B. S. Haldane, whose *Daedalus* of 1924 had offered an even more ambitious
vision of how science would transform human life. When we take into account
the numerous magazines such as *Popular Mechanics* promoting the technical
developments that would make life easier, we come to appreciate how a focus
on the work of literary figures can bias our view of how ordinary people at the
time thought about the future. Popular science writers often looked to the future
with optimism, and for many readers they were the most trusted guides.

So, was the early twentieth century a morbid age terrified of future wars or
a streamlined era fascinated with speed and convenience? In fact, it was
a complex mixture of both attitudes. In his study of the years preceding the
Great War, Philip Blom notes that many of the cultural, scientific and technical
innovations that galvanized society had their origins then. The title of his book,
The Vertigo Years, encapsulates a growing sense that the world was moving
into an age of rapid and unpredictable change which some found frightening,
but many experienced with exhilaration. Exploring the hopes and fears generated in France, Roxanne Panchasi writes of a 'culture of anticipation'.[7]

It all depends on where you focus your attention. The situation changed
decade by decade, and different social groups experienced the changes in
different ways. National experiences differed too. The post-war experiences
of the European nations were hugely influenced by ideologies, both of the right

and the left, which were enthusiastic for change. Conservative social and political forces viewed the new developments with suspicion. Nor did the divisions abate with time. By the late 1930s, Europe was lurching toward war, while 'streamlined' America was recovering its confidence after the Great Depression.[8] New tensions subsequently emerged in the era of the Cold War.

The image historians create depends on the material they study, and unfortunately most academics specialize in a particular period, nation, class or literary genre and present an image of the past that generalizes from their chosen area. We have thus accumulated a mass of secondary literature that presents a host of conflicting interpretations of early twentieth-century culture. It's a bit like the story of the blind men studying an elephant – each feels a different part of the animal and thinks that his impressions tell him what the whole beast must be like. We need to recognize that there wasn't a single coherent Western culture that responded uniformly to the prospect of scientific and technical progress. The time is ripe for a more comprehensive overview that will balance the pessimistic with the optimistic, the technophobe with the technophile, the warmonger with the industrial designer, the literary intellectual with the inventor. And the first step must be to gain a better view of just who was involved with predicting and shaping the future – and who was trying to sell the rival visions to the public.

Science Fact and Science Fiction

The evidence a historian uses may open a restricted window on what is actually a very complex situation. Literary scholars will tend to focus on novels and are most likely to prefer the highbrow literature produced by intellectuals. Students of popular culture might look to the more everyday novels – a very different genre, but one still produced mostly by authors with little experience of science and technical innovation. But if they turn to books and magazines dealing with popular science, they encounter material written by authors with real experience of science and engineering. These authors would tend to support the development of science and technology and would look forward to the next stages in the progress. Occasionally, a high-profile scientist such as Haldane would contribute to this literature, attracting newspaper headlines and arousing the fears of highbrow novelists and intellectuals.

The relationship between popular writing about science and science fiction can be quite complex and, in some cases, there is little clear distinction between the two areas. 'Science fiction' may be taken to include everything from the pulp magazines to the occasional ventures of literary intellectuals into the field of future dystopia. Our concern here is, of course, with efforts to imagine a future world – stories about invading aliens or colliding comets are not relevant unless new technologies are predicted to deal with them. But the aims of those who set stories in imagined future worlds can be quite different. Literary figures and moralists usually agonized over the potential threats to traditional values, while pulp science fiction authors worked hand in glove with the enthusiasts who promoted the latest developments.

Even those readers who do not enjoy the genre as a whole will be familiar with the future worlds created by authors such as Aldous Huxley and George Orwell. They are the subjects of biographies and literary analyses, and their works are routinely reprinted with scholarly introductions.[9] Their visions are almost all dystopias – nightmare stories set in a dehumanized world, with technological developments being depicted as the tools of regimes that control every aspect of life. Fritz Lang's influential 1926 movie *Metropolis* depicted the workers enslaved in the bowels of a futuristic megalopolis. French intellectuals were also suspicious of the drive toward a mechanized world and frequently identified America as the source of the trend. Georges Duhamel's *Scènes de la vie futur* of 1931 was translated as *America the Menace*. A variant on this theme is E. M. Forster's story 'The Machine Stops' of 1909, in which people have willingly adopted a life of ease in mechanized cocoons.

The extent of science's involvement in these stories varies, however. In the case of Huxley's *Brave New World*, it is considerable, but it has been argued that Orwell's *1984* is not a dystopia in the same vein because new technology plays only a limited role in how the state's control is exercised.[10] Orwell's target was not the threat of new technology, but the possible emergence of totalitarian regimes that would misuse whatever was available. It is fair to say that the same concern was paramount in virtually every dystopia from Owen Gregory's *Meccania* of 1918 through to Yevgeny Zamyatin's *We* of 1924 to David Karp's *One* of 1953. But for most of these authors, the power given to the state by new technology plays a greater role than it does in *1984* and they are thus of direct relevance when it comes to assessing visions of future developments.

Mainstream contributors to science fiction such as H. G. Wells, Isaac Asimov, Arthur C. Clarke and Robert Heinlein now attract a good deal of scholarly attention, although they seldom figure in the realm of literary criticism. Scholars are recognizing that as a window on popular culture, this literature is a much better guide than the highbrow novels that achieve iconic status only after they have been incorporated into the literary canon.[11] We also have studies on the spread of science fiction into the realm of cinema, and a study by Christopher Frayling notes how the technology imagined in the movies can actually inspire the scientist and engineers to create it.[12] There is also a mass of information and comment generated by fans, much now available via the internet.

These authors wrote 'hard' science fiction in which the effort to predict future technologies and their implications was a major inspiration. Their protagonists may engage in struggles against oppressive states (as in Wells's *Sleeper Awakes*), but they operate in a world which is very much shaped by new technologies. Asimov defined what he called 'real science fiction' as 'those stories that deal with scientific ideas and their impact in the future as written by someone knowledgeable in science'.[13] He points to Heinlein's early stories such as 'Solution Unsatisfactory' of 1941 which imagine worlds transformed by new technologies (in this case nuclear weapons) and place their heroes in situations defined by the problems created by the technologies. Heinlein himself preferred the term 'speculative fiction' and later writers such as Brian Aldiss have protested against the focus on the impact of science and technology. But it is precisely the kind of story identified by Asimov that resonates with the more sober futurology that he (and many others) also engaged in.

Along the same lines, Arthur C. Clarke argued that '*only* readers or writers of science fiction are really competent to discuss the possibilities of the future'.[14] What he meant was not that only science fiction can predict effectively, but that meaningful futurology has to involve an element of imagination. It cannot be mere extrapolation from existing trends – it has to involve choosing a conceivable technology and predicting what might happen if and when it is developed and applied. So even an account of a future world that does not include a human-interest narrative involves thinking about possibilities, but not certainties. The authors may get the details wrong when viewed in hindsight, but they tell us much about what educated people thought was at least plausible at the time. These stories are anything but utopias, but they recognize that scientific progress will continue and seek to grapple with the consequences. A partially dystopian scenario is, in any case, far more useful as a literary device than the perfect society envisaged by the real technophiles. It would be hard to set an exciting story in a utopia where everyone was genuinely happy and fulfilled all of the time.

There is also, of course, the 'pulp' science fiction of the popular magazines that began to appear from the 1920s onwards. These are usually dismissed as being of poor literary quality, and at the lowest level these stories offer only an impoverished vision of the future, even when major new technologies are imagined. Ray guns and spaceships replace six-shooters and stagecoaches, aliens replace Indians, but the stories often merely rehash the themes that would be familiar to any reader of popular westerns. There is little effort to develop a complex human story, but also little effort to think seriously about the effects that the new technical developments might have on society. Even so, the pulp fiction helps us to understand what some ordinary readers were prepared to accept as plausible visions of a future world. Just to imagine the possibility of space travel was a major imaginative leap until rockets transformed the situation in World War II. Before that, most would scoff at the prospect, but a sub-culture was building up which was more aware of the possibilities. By the time authors such as Asimov and Heinlein entered the field around 1940, the magazines had matured into a major source of inspiration for a generation of enthusiasts.[15]

The emergence of popular science fiction magazines reminds us that we need to take account of ordinary readers' interests as well as those of the literary intellectuals. Students of popular culture are well aware that popular science writing flourished in the Victorian era, but are taking note of important transformations in the early twentieth century. Books and magazines were increasingly well illustrated and there was considerable interest among some sections of the public in well-presented information about the latest developments. Newspapers, too, trumpeted the achievements of aviation pioneers and inventors of all sorts. Eventually new media such as film, radio and later television became active, opening up new avenues for those involved with science to educate the public and seek to influence its attitudes.[16]

Magazines such as *Popular Mechanics* in America and *Armchair Science* in Britain were dedicated to the promotion of technical innovations, and by their very nature stories on such themes invited speculation about future applications and their impact. The magazines also included occasional ventures into more far-reaching futurology, the predictions sometimes featuring in their cover illustrations (see colour plate 5). There was also a constant flow of books predicting the future of science and technology. Some were written by major figures in science, such as Haldane's *Daedalus* and J. D. Bernal's *The World, the Flesh and the Devil*. There were many similar works by lesser-known figures, most of whom had at least some technical education or experience. Historians of science now take popular science seriously as a means of accessing the public's attitude toward science, and if the relevant books and magazines were routinely speculating about the future, they must have played a role in shaping attitudes along with the fictional accounts.

It would thus be a mistake to draw too sharp a line between science fiction and the kind of futurology based on extrapolating from the current state of science. The 'harder' kind of science fiction often places its protagonists in a future world whose technological hardware has been conceived using the same predictive insights as those used by futurologists. A few novels are almost hybrids between the two genres, the human-interest narrative playing a very minor role against the background in which the author tries to imagine a future society. Wells's *The Shape of Things to Come* would certainly fit into this category, and the narrative element in Olaf Stapledon's *Last and First Men* is so thin that it scarcely classifies as a novel. Some authors wrote in both categories, including Wells himself, and later figures such as Isaac Asimov and Arthur C. Clarke.[17] 'Professor' A. M. Low, science consultant and later editor of *Armchair Science*, wrote futurological articles in the magazine, two books on the theme and also science fiction novels. Futurological texts were occasionally presented in the form of a fictional account of life in the predicted future.[18] The rhetoric of much futurological literature shows a remarkable similarity to that of science fiction, suggesting that the general public would regard them as parallel routes to the imagined future. Even the illustrations – including the magazine covers – look the same.

Two Cultures?

There was thus a complex of differing attitudes toward science and technology among those who sought to influence public opinion through the mass media. Writers of hard science fiction worked hand-in-hand with popular science writers to promote the hope of technical progress and explore its social implications. Literary figures were less interested in the technicalities and more fearful of the consequences. Wells pointed to the difference between the two outlooks in his 'Discovery of the Future' lecture of 1902, contrasting what he called the legalistic or past-regarding mentality of the majority with the creative or future-regarding approach of those who looked toward and wel-comed the introduction of new technologies and new social arrangements.[19]

The polarization Wells was trying to identify looks remarkably like an anticipation of the 'two cultures' model popularized by C. P. Snow in 1959,[20] in which society is divided between the humanities and the sciences. The humanities have a stranglehold on government which blocks technical research and development. Snow is now a much-maligned figure, his novels dismissed as second-rate and his views on the cultural divide criticized as over-simplified and self-serving. But I have a soft spot for him, not least because he was an old boy of my own school (alongside my future father-in-law). Historians have shown that his view of the humanities' influence was grossly exaggerated – the British Government had been investing heavily in scientific research for decades in the hope of creating high-tech weapons systems. But Snow was reflecting a view already developed by Wells, suggesting that there really was a sense among the technophiles that they were struggling against an entrenched attitude suspicious of their aspirations.[21]

Snow himself admitted that he simplified his images of the two cultures to make his point, and that his vision was more appropriate to Britain than to many other national contexts. But if we generalize the image that Wells and Snow projected, we get a useful handle on the different perspectives from which people viewed the prospect of scientific and technical progress. On the one side, we have those directly involved with scientific and technical work, and on the other, those who reflect and comment on the human situation and social issues. This division obviously misses out a huge swathe of the population whose only concern is making a living or a profit within the existing system – they are the ones whose opinions might be influenced by writing about science or technology, either factual or fictional. There are also people active in areas such as economics and politics who hope to influence things by proposals that may not involve technical or industrial research.

The literary figures and highbrow intellectuals represented the humanities side of Snow's divide and the legalistic side of Wells's. Their views would be shared by a wider swathe of popular novelists and often by the journalists who reported and commented on social and political news. The key point is that these would be people with little or no education in the sciences and few contacts with anyone involved with scientific, technical or industrial work. For this reason, the authors of hard science fiction would have to be excluded from the group. Snow dismissed the literary intellectuals as natural Luddites who instinctively feared that the new gadgets produced by technical research would be misused either by the military or by political demagogues. The negative viewpoint of major literary works such as *Brave New World* suggests that there may be something in this characterization. But so does a whole genre of more popular stories in which scientists are depicted as lone madmen and future wars rendered horrific by new weapons.[22] Journalists might be equally biased – there were few dedicated science correspondents until the middle decades of the century and the ordinary hack would write the occasional 'gee whiz' story about new technology that might promote either wonder or ridicule.

If social commentators were instinctively worried about the effects of scientific and industrial progress, Snow claimed that the scientists had the future in their bones. Their work depended on the discovery and application of new knowledge and even when well aware of potential military applications, they were supportive of technical progress. Snow tried to focus on academic science, but as a technocrat he knew that most research was now being done in government or industrial laboratories. He also ignored the amateur inventors – the archetypical lone researchers, now increasingly overtaken by 'big science', but not yet extinct. For Americans, the role model here was Thomas Edison, and pioneers such as Guglielmo Marconi and John Logie Baird showed that such figures could still play a role. These figures interacted with engineers, designers, architects, progressive industrialists and a host of men (almost always men) with significant technical expertise derived from work with industry or the military. Some politicians, often of the far right or left, also promoted the role of scientific research in the hope of national glory or social progress.

Snow conceded elsewhere that the scientists themselves were often 'gadge-teers' obsessed with perfecting the technique in hand, unable to see the broader implications of what they were doing.[23] They might promote the potential value of their work, but in a short-sighted way – and much of the popular futurology written by technical experts tended to envision possible technologies without thinking through the problems they might create in practice. But the vision displayed by figures such as Haldane and Bernal suggests that, at their best, the research scientists could imagine the broader sweep of what new discoveries might achieve.

We are often told that the scientists and technical experts didn't communicate with the public because they were too busy or feared it would damage their reputation. But as I showed in my *Science for All*, this image is misleading. Large numbers of them tried their hand at popular science writing from time to time and a few developed real skills in the field and earned part of their reputation (and some of their income) from it. For every big name such as Haldane who participated, there was a host of lesser-known figures contributing regularly to science magazines and writing books in the field. A. M. Low provides a classic example of someone who combined the careers of inventor and science writer. These people interacted with the hard science fiction writers (and sometimes joined their ranks) to provide a source for the public not only of technical information, but about the prospects for future developments. Their contributions certainly did not ignore problems such as the emergence of new military technologies, but were generally enthusiastic about the future benefits that could be expected.

Historians and the Idea of Progress

We are now in a better position to address the question raised at the start of this chapter: why do scholars offer such different impressions of early twentieth-century attitudes toward the future? The answer lies in the sources from which they derive their evidence, because different sources reflect the views of different communities. The literary scholar focusing on highbrow novelists is much more likely to encounter pessimistic views about the future development of science and technology than the historian of popular science. An analysis based on political and economic journalism will not look the same as one derived from pulp science fiction. If we are to get an overview of what the reading (and listening) public was presented with by the mass media, we must take all of these sources into account and be prepared to accept that views on the future of science will form a complex and ever-changing mixture. The mix will reflect the professional background and nationality of those who write and broadcast and will change decade by decade if not year by year.

An example of how perspective can be shaped by sources can be seen in the work of two authors already identified as depicting the interwar years as a period of despair, overwhelmed by the horror of war, economic depression and oncoming totalitarianism. I. F. Clarke's *Pattern of Expectation* has a series of chapters on this period culminating in chapter 9 entitled 'From Bad to Worse'. Richard Overy drives the point home by calling his analysis of the period *The Morbid Age*. Clarke's sources are almost all novels, mostly highbrow dystopias and popular novels predicting war.[24] Overy ranges more widely, but much of his source material is derived from the writings of intellectuals and political commentators who had little knowledge of or sympathy for science. As one reviewer of Overy's book pointed out, his analysis misses out the literature aimed at an increasingly large and still prosperous middle class that welcomed new technologies because they were transforming life for the better.[25]

Chapter 9 of Clarke's *Pattern of Expectation* describes twenty-six novels, all but two of which are pessimistic, including eight published in Britain as war loomed in 1939. Yet buried in the chapter is a reference to endnote 21, which lists forty-one additional books, mostly non-fiction and mostly optimistic about the future. None of these is mentioned in the text or used as evidence to seek a more balanced perspective. Ten of them are predictions of future scientific and technical progress. My own work has expanded this list to include over twenty book-length futurologies published between the wars, all of which are either optimistic or at least present a balanced assessment.[26] All are by authors familiar with science or involved in one way or another with scientific, technical, industrial or medical progress. This list does not include works by H. G. Wells (who remained optimistic about the long-term future until the late 1930s) or the group of short but influential books published in Kegan Paul's 'Today and Tomorrow' series in the 1920s.[27]

Overy explicitly set out to counter what he saw as a false image of British optimism between the wars presented in books such as Martin Pugh's *We Danced All Night*.[28] But the pessimistic view reflected in his choice of sources – and even more so in Clarke's – is similarly one-sided. This interpretation actually reflects the more common interpretation of Western culture as having lost its faith in the idea of progress in the face of the disaster of the Great War and the economic depression of the 1930s. This was the age in which Oswald Spengler's *Decline of the West* became a best-seller. No one could deny that confidence was shaken, but there is also plenty of evidence to suggest that the atmosphere was not one of permanent doom and gloom. Both the war and the financial crash were seen as evidence that the Industrial Revolution of the previous century had been mismanaged – but the Marxists were not the only enthusiasts who hoped to reform the economy through rational planning. This was Wells's approach, although he (unlike many economists) recognized that the future would be shaped as much by new techniques as by new methods of distribution.[29]

We also tend to forget that religion still played a significant role, though declining at least in Europe. Liberal religious thinkers continued to call for moral and social reform, seeing progress as the unfolding of God's plan for the world.[30] This approach was also taken over by humanists such as Julian Huxley and incorporated into grandiose plans for world unity.[31] Even Sigmund Freud, not noted for his optimism on the human situation, offered the hope that eventually traditional religion would be transformed.[32]

Julian Huxley's role in these activities reminds us that the theory of evolution was still seen by many biologists as a valid foundation for a progressionist world view. The synthesis of Darwinism and genetics revitalized evolutionary biology without destroying the hope that the overall trend was still upward. The new theory was not widely discussed in the popular press, but major fossil discoveries were covered and were usually presented in a way that emphasized the overall advance of life. The huge publicity surrounding discoveries of prehistoric human remains (including the Piltdown forgery) kept the idea of long-range social and technological progress before the public. The rising tide of Fundamentalist Christianity in America only partially offset this programme, and was not duplicated in Europe (where liberal religion was challenged more effectively by the neo-conservatism reflected in the writings of C. S. Lewis).[33] For many readers of the popular press, the claim that human progress continued the upward path of evolution remained plausible and thus offered hope for the future even if the immediate prospects were daunting.

National differences are important here as in many other areas. The Great War affected European nations far more than it did the United States, but the 1920s were the 'roaring twenties' on both sides of the Atlantic. France in particular worried about the threat of Americanization. The economic depression hit America harder than Britain, but by then the lives of ordinary people had already begun to be transformed by new technologies, including radio, the automobile and a flood of household gadgets.[34] By the late 1930s, America had regained its confidence, as reflected in the enthusiasm for streamlined designs and Futurama display at the New York World's Fair of 1939–40 (Figure 1.1; see also colour plate 2). When that exposition closed, Europe was already engaged in the war that many had seen approaching for several years. Those fears certainly account for the rise of pessimism in Britain and France, but we should not forget that in Russia, Germany and Italy totalitarian regimes of the Left and the Right proclaimed the march of progress and often saw technology as the key to the future prosperity offered to their people.

The survey offered here cannot hope to cover visions of the future across the whole of the Western world in the early twentieth century. It will focus on America and Britain, with references to developments in continental Europe, especially where they hit the headlines. But within this limited remit I hope to include material from a wide range of sources that will give us an overview, even if a kaleidoscopic one, of the ever-changing debate about future prospects. Books and magazines will form the main source of information, with inputs from newspapers, radio, film and eventually television. No one could hope to

Fig. 1.1. The Futurama exhibition at the 1939 New York World's Fair. Spectators view a futuristic cityscape from moving chairs. From Wohl, *Spectacle of Flight* (25), p. 297

exhaust the whole range of material potentially available, even with modern techniques of access and analysis, but I have tried to gather representative samples from the more prolific sources and have depended on the secondary literature, both academic and popular, for inspiration and guidance.

Getting It Right?

One last point should be made before getting down to details. We should not approach this material from the past in order to judge the accuracy or otherwise of the predictions made in it. It is all too easy to wonder at the visions of a future metropolis on the cover of a 1920s magazine, or sneer at some predicted technology that never came to fruition. To be sure, there was a certain amount of hand-waving used to invoke new scientific principles for which there was no justification beyond the free exercise of the imagination. Wells engaged in this kind of speculation when he imagined the gravity-blocking material that drove the space vehicle in his *First Men in the Moon*. It was all too easy for an author unfamiliar with what was going on in science to invoke something out of the blue, especially at a time when the most popular image of the scientist was that of the mad lone inventor. But even here we should not, perhaps, be too judgmental – after all, eminent physicists dismissed powered flight as impossible, and no one predicted the huge impact of the personal computer in the real world. Most predictions (including those of Wells himself) rested on at least some foundation in the exploratory science and technology of the day and extrapolated from an element in the ferment of research. In some cases, the idea for a new technology was sparked by an innovation in pure science, but remained unrealizable for decades – atomic power is an obvious example. Even where new technologies were actually emerging, we are learning from the history of technology to be cautious about the application of hindsight.

The problem is that once a technology has become successful, we tend to assume that it had a clear-cut superiority over its rivals and that the superiority should have been obvious to anyone at the time. Surely, electricity provided better lighting than gas, or aeroplanes a better mode of transport than airships. Yet historians of technology studying episodes where rival systems battled for supremacy in the marketplace find that the outcome was never obvious to those actually engaged in the struggle at the time.[35] A well-established technology can sometimes block the introduction of another that might well have succeeded had it got there first. Something that seems plausible in principle might have practical implications that block its introduction, and we seem poor at anticipating these consequences. Once powered flight became a reality, there were many who hoped that ordinary people would be able to operate their own personal aircraft – but few thought about the problem of air-traffic control around a big city. It was the suburban railroad and the motor car that made the projected megalopolis of Wells's *The Sleeper Awakes* less plausible, but raised the question of whether the suburbs should be replaced by garden cities.

Once the wider social and commercial factors are introduced into the equation, predicting the next successful technology becomes difficult, if not impossible. This is why Arthur C. Clarke suggested that only those familiar with science fiction are competent to predict the future – they are instinctively aware of the fact that they are picking out one possibility from many that are equally conceivable. They are, in effect, predicting the unpredictable.

Another complicating factor is the tendency to assume that each new technology has a single inventor. Indeed, the British patent system (unlike that in the United States) is based on the assumption that a single innovator can be identified.[36] In a world where it was assumed that scientists and inventors were always isolated individuals, it seemed obvious that new technologies had easily identifiable origins, which in turn encouraged the belief that by recognizing a promising inventor one could predict what might emerge. In fact, new technologies were increasingly the products of large research and development teams. But even where single individuals were recognized as the key innovators, it was almost invariably the case that to get a new system into commercial operation required a sequence of inventions, all of which had to be integrated even if made by separate individuals. Marconi became recognized as the 'inventor' of radio, but had to compete with other interests (including those of the physicist Oliver Lodge) which he had either to eliminate or to absorb. The complexity of the whole process limits the ability of anyone to predict in advance which technologies will succeed.

The problem of prediction becomes all the more serious the further one speculates into the future. Many of the examples studied below represent quite short-term predictions based on a design, invention or prototype already available at the time, and even here the level of accuracy in the predictions was remarkably low. Trying to imagine a future world that works with a whole host of technologies beyond those currently available is virtually impossible because we simply cannot foresee the interactions between the technically possible innovations, let alone the social impact of those that actually catch on in practice. Yet eminent scientists such as Haldane and Bernal were keen to explore the wider implications of new developments that were only just becoming visible even in principle. Their opponents naturally reacted when the predictions involved threats to the social order or to the very fabric of human nature.

The fascinating point of the whole exercise for the historian is that people were so keen to make the predictions even when it ought to have become obvious that they were unlikely to be successful. The motivation for those who were trying to promote a new technology is obvious, but the whole enterprise goes far beyond the cut-and-thrust of commercial competition and suggests that we do have a genuine fascination with the future – whether we welcome or fear what it may bring. The early twentieth century marks the point at which ordinary people began to take for granted that the future would be different and to think seriously about what it would bring. We may learn something of value from their efforts, if only the need to take current promises and threats with a pinch of salt.

2　The Prophets

Their Backgrounds and Ambitions

The future was contested territory. Edward Bellamy's *Looking Backward* of 1888 precipitated an explosion of interest in exploring the possibility of future social developments. But Bellamy's vision of a socialist utopia in the year 2000 focused almost entirely on the possibility of moral and political reform. Technological improvements were mentioned only peripherally. J. J. Astor's *A Journey in Other Worlds* of 1894 depicted an interplanetary voyage in the year 2000 from an Earth transformed by the applications of electrical power. Astor provided some details of the new technologies he envisaged, including a plan to 'straighten' the Earth's axis to improve the climate. But, again, the main purpose of his book was to use the idea that the other planets were inhabited to convey a message of moral and, in this case, spiritual significance. For an increasing number of writers, however, it was technology itself that would promote the transformation of society, for good or ill.

There had been some predictions of future technical developments earlier in the nineteenth century. Jules Verne highlighted the possibility of submarines and aircraft long before they were practical – although his stories take place in the present and depend on visionary inventors working in secret. Winwood Reade's *Martyrdom of Man* had concluded by proposing three inventions that would transform the human situation: better sources of power, aerial locomotion and artificial food. By the turn of the century, there was a growing confidence that major advances would soon be made and would have an immense impact on society. There were 'technological utopians' in America who foresaw a society transformed by air travel, giant cities and the like.

New discoveries in science and new technologies were now overthrowing old certainties and prejudices. Aviation, long dismissed as a fantasy, suddenly became a reality. Radio, X-rays and radioactivity showed that science could provide knowledge capable of transforming our lives, as well as our view of the world. The expansion of popular publishing, especially in the realm of newspapers and magazines, created opportunities for the promotion of these new ideas and for speculating about further advances. Nikola Tesla temporarily gained a reputation nearly as great as Edison's through his promises of new ways of applying and distributing electricity.[1]

The writings of H. G. Wells at the turn of the century made the development of science and industry central to a projection of social transformation. He realized that any prediction about how society might change had to include not only political events, but also the impact of new technologies that would transform everyday life. Focusing on a particular technology such as aviation could suggest that even a single innovation could be transformative. But a mass of innovations could interact to create a whole new world. Whether that world would be a nightmare or a utopia would depend on society's response.

As interest in the opportunities created by science and technology grew, lobbyists from various backgrounds outlined in the previous chapter sought to convince the public that their own visions were the most perceptive. For some intellectuals, the most significant impact of scientific progress lay in the unsettling philosophical implications of the new physics. But the main thrust of futurological speculation arose from the belief that applied science would transform society and even human nature. Enthusiasts for the new technologies realized that they had the potential to revolutionize daily life and the more adventurous saw that the long-term effects would be even more transformative than any revolution at the purely philosophical level. It was precisely these long-term consequences that so disturbed conservative thinkers anxious to defend traditional values.

H. G. Wells straddled the various genres of publication and anticipated most of the alternative ideologies. He wrote popular novels and campaigned endlessly to reform society along lines that would make the future better for all. He recognized that new weapons such as aircraft would transform war and make it far more destructive. He also agreed with those who feared that if there was no social revolution to redistribute the fruits of industry we might be driven into a world of ever-increasing inequality where science would be misused by the rulers. Yet through most of his life he retained the hope that humanity could transform itself through rational planning so that social and technological progress would accelerate into an ever-brighter future.

Wells personified the division emerging between those who looked toward the future with optimism and those who feared the worst. Technophiles were by no means blind to the military applications, but saw the benefits for everyday life outweighing the threats. Pessimists often had no training in science and no experience with industry, although a few did have access to detailed information about science – Aldous Huxley's brother Julian Huxley was a well-known biologist who worked with Wells. They feared the application of science, but couldn't agree on whether the greatest threat came from its misuse by the military or by a totalitarian

dictatorship. In the latter case, Wells's own vision of a rationally ordered world became the nightmare.

Rival perspectives were presented to the public in a variety of formats ranging from highbrow novels to popular articles in magazines and newspapers. Those interested in science and technology would read popular science magazines and science fiction to reinforce their perception that the future was likely to bring progress. Those with literary or political interests were more likely to read novels and magazines reflecting the pessimistic viewpoint of their peers. Newspapers and general magazines were contested territory and reflected fragmented impressions challenging for the historian trying to gain a coherent overview. By the 1920s, many newspapers had columnists writing regularly on aviation, cinema, motoring and some even on science itself. Their authorities also wrote for the popular science periodicals and tended to reflect a positive vision of what was likely to become available in the near future. Newspapers might also carry reports of politicians worrying about attack from the air or speeches by literary figures lamenting the destruction of the countryside by motorists. Political affiliation was no guarantee of consistency – even left-wing papers knew that their readers went to the cinema and aspired to own a car. The *Daily Herald*, which commissioned Ritchie Calder to write the articles which became his *The Birth of the Future*, supported the Labour Party.

The Shape of Things to Come

Wells is best remembered as a novelist, especially for his science fiction, but he was trained in science and remained committed to the popular dissemination of scientific knowledge. His speculations about the future came from a familiarity with current developments in science and reflected his expectation that its practical applications would help to shape how we will live. At the same time, he was bitterly critical of an economic and social system that left ordinary people cut off from many of the potential benefits. For Wells, the technical experts who were transforming industries and creating new ones had the organizational skills needed to create a rational society in which all would find a meaningful place. His critics feared that they would be forced to take that place whether they liked it or not.

Wells gained a scholarship to the Normal School of Science in London in 1884. He left without a degree to work in science journalism and write textbooks, gradually expanding his sphere to place stories in popular magazines. Many of these had scientific themes, culminating with 'The Time Machine' in 1895, which made his reputation. He remained active in the field of educational publishing and enjoyed a huge success with his *Outline of History*, published

Fig. 2.1. H. G. Wells, his son G. P. Wells and Julian Huxley discuss the
production of their popular work *The Science of Life* (1929). From Julian
Huxley, *Memories*, between pages 126–7.

originally in serial parts in 1919. This concluded with a call for the formation of
a world government as the only way of guaranteeing progress following the
disaster of the Great War. *The Science of Life*, written in conjunction with Julian
Huxley, was a follow-up to the *Outline of History* and contained one of the first
popular accounts of the emerging synthesis of Darwinism and genetics
(Figure 2.1).[2]

'The Time Machine' passed over the near future to deposit its hero in an age
when the human race had split into two species descended from the wealthy
classes and the workers. In 1899, the polarized society in which this speciation
might begin was imagined in 'A Story of the Days to Come' and then in the
novel *When the Sleeper Awakes*. As Wells later confessed, this was 'essentially
an exaggeration of contemporary tendencies: higher buildings, bigger towns,
wickeder capitalists and labour more downtrodden than ever and more
desperate'.[3] Everyone now lived in huge roofed-over super-cities, travelling
on 'moving ways' and powered by wind-vanes. Aircraft were used for long-
distance transport and for war. The huge scale of the book's vision of the future
impressed many – as George Orwell later said, no one ever forgot their first
reading of it.[4]

Wells also proposed some serious predictions about what would happen in
the next century for the *Fortnightly Review*, published almost immediately in

book form as *Anticipations*. Wells later claimed that this was 'the first attempt to forecast the human future as a whole and to estimate the relative power of this and that great systems of influence'.[5] Here was serious futurology unencumbered by the need for a plot, developing both ideas about the effects of new technologies and call for social revolution. He already realized that the development of suburban railways and better roads would prevent the over-concentration of population in mega-cities. New devices in the home would transform domestic life, with most people living in apartments and sending their children away for education. Flying would come, but only slowly – as he confessed in 1915 his prediction that effective aircraft might be available by 1950 was hopelessly pessimistic.[6]

The book's real impact, however, was in its analysis of the dangers of capitalism and the weakness of democracy to transform the system. Wells called for a 'new republic' governed by experts who would maximize production and ensure fair distribution. The harsher side of his thought was all too apparent, though: 'The men of the New Republic will not be squeamish, either in facing or inflicting death, because they will have a fuller sense of the possibilities of life than we possess.' The unfit would be painlessly eliminated, the mentally ill would be encouraged to suicide out of a sense of duty and the inferior races of the world would face extinction.[7] It was this aspect of Wells's social programme, developed through the rest of his career in works such as his *A Modern Utopia* of 1917, that so alarmed some contemporaries. It was not so much the direct effects of new technologies they feared as the ruthlessness of the rational experts who might use those technologies to reshape the world.

In 1908, Wells expanded his prediction about the coming of aviation in *The War in the Air*. A classic 'future war' novel, this predicted a global conflict sparked by the attack of a German air fleet on Britain and America. Curiously, the setting is a world in which ground transport has already been revolutionized by gyroscopically-stabilized monorails.[8] Butteridge, the lone inventor of a superior design of airship, sells it to the Germans, who build their fleet in secret and spring a surprise attack. Small airplanes are also in use, but the main fighting fleets are composed of airships. In the end, all major cities are destroyed and life continues only in rural areas.

Wells's imagination ranged even further in *The World Set Free* of 1914, dedicated to Frederick Soddy for his suggestion that radioactivity might offer a new source of power. In the novel, the scientist Holsten discovers how to produce energy from radioactive materials in 1933, just in time to prevent the world's industries running out of coal and oil. The later chapters return to Wells's political ambitions, describing a war caused by the failure of society to reform itself in parallel with the developments in technology. A 'Modern State' ruled by technocrats emerges from the chaos following the war. The novel ends with the reflections of the dying leader Marcus Karenin: 'Science is no longer our servant. We know it for something greater than our little individual selves. It is the awakening mind of the race.'[9]

By the early 1930s, Wells was still confident that progress would continue 'despite human fear and folly', as the conclusion to his *The Work, Wealth and Happiness of Mankind* put it.[10] But he retained the idea that first the old social order would have to be swept away. The theme of a conflict that comes close to destroying civilization was the starting point for his most ambitious book about the future, *The Shape of Things to Come*, published in 1933 and filmed two years later by Alexander Korda (see colour plate 1). Presented as a history written in the future and somehow transmitted back to the present, the book alternates stories set in particular periods with mock historical surveys. In the coming war, Britain and most of the civilized world is reduced to barbarism by the effects of bombing. Eventually, a new society created by the few surviving aviators spreads around the world, subduing the primitive tribes of Britain and elsewhere with soporific gas spread from the air. We now move to the new world order created on rational principles and run by technical experts, the 'scientific samurai', who eliminate their opponents and enforce conformist behaviour through 'educational control'.[11]

In the movie *Things to Come*, for which Wells acted as a consultant, the technocrats plan to launch a vessel to the moon from a giant 'space gun'. They are opposed by a conservative faction led by the sculptor Theotocopulos, who fears that we are being dehumanized by the drive to dominate nature. In the final scene, the leader of the samurai, Oswald Cabal, offers humanity the choice between stagnation and endless future progress, asking 'Which shall it be?'[12] The film was a box-office flop in America, but stills were widely published in the popular press and hardly anyone could have remained unaware of its message.

Wells noted that there was speculation in the 1930s that global aviation would require an international authority to control it – his fictional 'Transport Union' was merely an extension of the idea.[13] Another popular novelist who imagined that the requirements of air travel would entail the setting up of a global authority was Rudyard Kipling. In 1909, his 'With the Night Mail: A Story of 2000 A.D.' provided a vivid description of a world transformed by long-distance air travel, complete with fictional advertisements. Kipling had no scientific training, but he had worked with engineers during his time as a journalist in India and kept in touch with naval officers after his return to England. The story is set on a transatlantic airship which can travel at over 200 miles per hour thanks to a new technology, 'Fleury's ray'. The airships and their operations are described in some detail, reflecting Kipling's familiarity with shipping affairs. But he fudges the science, claiming that no one really knows how the ray works.[14] The real point of the story is that the new technology by its very nature will require new social structures. Already the Aerial Board of Control (ABC) which regulates the system has gained considerable power over national governments.

In 'As Easy as A.B.C.' – published three years later – Kipling takes his imagined world forward to the year 2065, by which time the ABC has become the chief global authority. Its power derives from the fact that personalized air transport has made society more individualistic by allowing people to live in isolated homes.[15] The population has been dramatically reduced and is still falling. The old political system with parliaments to represent 'the will of the people' is now an anathema to almost everyone. When a few 'servile' types try to reinstate the old ways, the ABC is called in to prevent the resulting outcry from turning violent. Unlike Wells, Kipling's ABC adopts a hands-off approach to world government, confining itself to keeping the airways open. Its motto is 'Transport is Civilization' and everyone can do what they want, as long as they do not interfere with the traffic 'and all it implies'.

Literary Dystopias

Wells's vision of an autocratic state run by experts appalled those who preferred to retain traditional values. The result was a series of books, plays and movies all designed to expose the harsher implications of a rationalized world order. Wells's own flirtations with totalitarian regimes of both the Left and the Right did not help matters. But it was the threat of authoritarianism itself that really frightened the intellectuals and moralists – new technologies were relevant only if they were perceived as crucial for providing the state with better means to control the population. Wells's technocratic samurai were one possible version of the imagined elite government, but the totalitarian states emerging in Russia, Italy and Germany offered models that did not depend on the scientists themselves taking over. Some of the future dystopias made little appeal to new technologies.[16]

A small number of stories imagined a dehumanized future into which the population has stumbled through apathy and laziness. Here, there was no rational and ruthless elite out to enslave us all, just the relentless march of machines offering an easy life cocooned from harsh realities. E. M. Forster's story 'The Machine Stops' appeared in the same year as 'With the Night Mail', offering a vision of the future in which people live isolated lives in underground cocoons, never venturing out and communicating only via electrical technologies. All their wants are supplied automatically and their lives are essentially passive and pointless. There is no suggestion that they have been compelled to live this way – it just became more convenient to escape the pressures of the outside world. Eventually, they begin to worship the machine that supplies all their wants as a god. But the designers who set the system up were evidently not rational enough, because it begins to break down and no one knows how to repair it. The underground population becomes extinct, although a few primitives survive in the outside world to continue the race.[17]

A more sinister version of the same threat appears in Olaf Stapledon's *Star Maker* of 1937. Here, an interstellar traveller visits an alien race, the 'other men' who have advanced radio technology. They develop portable receivers capable of direct stimulation of the brain to produce any sensation, including sex. People soon begin to spend their whole life in bed, cared for by automatic machinery and deriving all their experiences from the radio network. Reproduction is by ectogenetic processes. In Stapledon's alien world, however, there are dark forces at work in the form of commercial interests which promote the new system, but eventually realize that they will soon have nothing left worth controlling. They take over the radio network to broadcast nationalistic propaganda, leading to the outbreak of a war that more or less destroys civilization.[18] As with his *Last and First Men* (discussed below), Stapledon ends up warning against the threat of war, but his vision of a race that willingly enslaves itself through passive entertainment offers a significant twist on the theme.

Wells's early vision of a working class enslaved in the bowels of a vast mega-city found its way into the popular imagination, most obviously in Fritz Lang's 1926 movie *Metropolis* (although much of the film's action is Gothic melodrama).[19] The most striking dystopias of the period are those depicting the human race of the future as dehumanized by totalitarian regimes that impose a totally regimented lifestyle. But here the threat was political or ideological in a broader sense, the regime being either bent on acquiring power for its own sake or dedicated to a purely rational or managerial approach to the social order. One typical dystopia, Owen Gregory's *Meccania* of 1918, made no mention of futuristic technology.

Much better known is Yevgeny Zamyatin's *We*, written in the early days of Soviet Russia and first published in an English translation in 1924. Zamyatin had edited Russian translations of works by Wells, and his totalitarian state of the twenty-sixth century clearly owes some of its inspiration to his vision. But it also reflects concerns about the managerial approach to social organization being adopted by the Soviet authorities.[20] The Soviets welcomed biological science as a vehicle that would help them transform humanity, and there were several authors who wrote fictional accounts of the potential results. These hopes were reduced to more utilitarian concerns in the Stalin era, and Zamyatin was already warning of the dangers that state control might get out of hand. The population of his imaginary state is rigidly controlled in all aspects of life, including sex, and individuals are known by numbers rather than names. The image of an enclosed city is certainly Wellsian and there are indications of technological advances – aircraft are in use and the narrator is the designer of a spaceship being built to spread the benefits of the rational order to other planets. The most threatening technique is an operation, presumably lobotomy, used to make individuals totally compliant.

For very different reasons, Wells and Zamyatin inspired a number of later dystopias. The best-known vision of a nightmare future was Aldous Huxley's *Brave New World*, published in 1932. The biologist J. B. S. Haldane had predicted that eventually the human race would reproduce artificially by ectogenesis. He was certainly known to Aldous because he had collaborated with his brother Julian Huxley on a biology textbook. Aldous had already caricatured him as the physiologist Shearwater in an earlier novel, *Antic Hay*. His visions extended the range of popular speculations about the power of medical technologies to transform the body for the good – although Huxley was by no means the only novelist to predict that things might go horribly wrong.[21]

Haldane's prediction of artificially produced babies provided the model for an imagined future in which the state modifies its people so that they are perfectly fitted to their occupation and social class and are thus happy in their servitude. In the Central London Hatchery, babies are produced in identical groups by dividing the fertilized ova and after birth are psychologically conditioned to fit their allotted position in the community. Everything is organized along lines prefigured in Henry Ford's assembly lines. The supply of unlimited sex and a hangover-free drug, soma, ensures contentment for everyone except Bernard Marx, for whom the prenatal conditioning seems to have gone wrong. Marx's involvement with the 'savage' – introduced into the Brave New World from a reservation where primitive tribes are kept to amuse the population – allows Huxley to contrast the soulless efficiency of the planners with the deeper concerns of the artist.[22]

The misuse of biology is also a theme in Karel Čapek's play *Rossum's Universal Robots*, published in 1920 and translated into English in 1923. These robots are not mechanical – they are artificially produced living things manufactured by a powerful commercial company for use as industrial and domestic slaves. When they are eventually given souls, they revolt and destroy the human race. Being artificial, they have not been given the power to reproduce, but at the end of the play an experimental programme is in place to create a robot Adam and Eve.

We and *Brave New World* both warned that a society based on rational planning would be soulless even if ostensibly claiming to promote happiness. The totalitarian societies that emerged in Germany and Russia highlighted the threat, and in the aftermath of World War II George Orwell's *1984* created perhaps the most disturbing image of a future world. Orwell had at first admired Wells, but now saw the dangers of his rationalist ideology. His real target, though, was the political ideologues who wish to organize society only so that they themselves can control it. The emergence of the new world order is described in passages from the anti-Party book by Emmanuel Goldstein reminiscent of the fake history texts used to predict the future in *The Shape of Things to Come*.[23]

Alexandra Aldridge suggests that *1984* is not really a scientific dystopia because advanced technology plays only a limited role in the state's manipulative system.[24] The hero, Winston Smith, lives in a London hardly changed from the austerities of 1948, apart from the huge buildings housing the ministries of the ruling Party. The three great powers of the world are in a constant state of war, which they use as an excuse to deny the population the fruits of industry. But the war is pretty conventional – atomic bombs are available, but are not used for fear of destabilizing the balance of power and the missiles that land on the city are just like the V-2s that London had endured in the real war. The only advanced technology used by the Party is the telescreen, which can spy on the viewer as well as broadcast propaganda – a fairly predictable extension of what was already available. The most frightening aspect of Orwell's story is that those in power can use conventional means of torture to force their victims to think along the lines they demand.

Orwell's nightmare vision of state control was by no means the last. C. S. Lewis's science fiction trilogy beginning with his *Out of the Silent Planet* of 1938 plays out a cosmic struggle against soulless materialism on an interplanetary stage. Although Lewis claimed to have enjoyed Wells's 'fantasies', his neo-conservative Christian world view inevitably led him to attack the ideology of rational progress.[25] David Karp's *One* of 1953 postulates a superficially benevolent totalitarian state which generates happiness by demanding conformity. In this case, artificial means are used to reshape the personality of those who object. By the 1950s, the huge expansion of science fiction publishing was creating a flood of futuristic visions, now increasingly pessimistic in outlook.

The Future War Novels

In *The War in the Air*, Wells predicted the devastating effects of aerial bombardment and *The World Set Free* imagined an atomic bomb that would render cities permanently uninhabitable.[26] *The Shape of Things to Come* suggested that a worldwide war would be needed to destroy the old social order and leave the way free for reconstruction. In 1930, Stapledon's *Last and First Men* predicted wars between the great powers in which lethal gases spread by aircraft would devastate whole territories.[27] These were the most high-profile attempts by literary figures to warn about the potentially catastrophic effects of new weapons created by applied science. A flood of novels by lesser writers made similar predictions, peaking in the late 1930s as Europe again lurched toward crisis. Some of these novels postulate no new advances in military technology, reflecting only the geopolitical tensions of the time.[28] But a surprising number followed Wells's lead by imagining wars made more devastating by new weapons.

American novelists were active in this field at the turn of the century. They imagined invasions from Europe or Asia, thwarted when American ingenuity creates new and more powerful weapons. Thomas Edison was often portrayed as the potential saviour of the nation – a picture he did little to disavow.[29] Roy Norton's *The Vanishing Fleets* of 1908 had 'radioplanes' – airships using an anti-gravity device powered by radioactivity – defeating a Japanese invasion. In a preface the author claimed that its 'scientific possibilities are endorsed by some of the best-known inventors in the world, and notably by one of the greatest, no less a personage than Hudson Maxim [sic]'.[30] Hollis Godfrey's *The Man Who Ended War* of 1910 had a lone inventor forcing the world to disarm by threatening to destroy all war fleets with a disintegrating ray. As American involvement in the Great War loomed, Germany became the chief threat. Cleveland Moffett's *The Conquest of America* imagined a German invasion in 1921 driven back when their fleet is destroyed by torpedoes launched from airships. Future war novels were less common in America after the war, although British naval expert Hector C. Bywater predicted that a Pacific war with Japan would occur in 1931, anticipating the attack on Pearl Harbor that would take place ten years later (although he thought it would be launched in the Philippines).[31]

British novelists also responded to the nation's growing fear of invasion from the air now that the Royal Navy could no longer prevent attack by sea. In the period from 1900 to the outbreak of World War II, over twenty authors contributed to the genre. There was widespread fear of aerial attack, which increased dramatically in the years immediately before the outbreak of war in 1939. It was widely believed that poison gas would be used with devastating results. Images of vast fleets of aircraft rendering whole cities uninhabitable played on genuine fears articulated in the press.[32] Harold Nicolson and J. B. Priestley were two well-known writers who followed Wells in postulating the development of an atomic bomb.[33] Those with wilder imaginations included death rays and even biological weapons in the repertoire of destruction.

Many of these writers are now hailed as pioneers of science fiction, although this was not as yet a recognized genre and their novels were usually advertised as 'imaginative' additions to publishers' regular lists. Unlike the pioneers of science fiction in America, most had no technical expertise and little enthusiasm for science and technology. A few did know what they were writing about. Neville Shute had a career in the aviation industry before he became a novelist and the Earl of Halsbury had worked in the Royal Air Force. Hector Bywater and E. F. Spanner were involved in the public debates over the use of air power and used their novels to drive home their warnings. But for the most part, these were professional authors with few scientific interests, some of whom achieved considerable popular acclaim. Michael Arlen was a hugely prolific novelist of the interwar years, while J. B. Priestley came to be regarded as a British national treasure. Harold Nicolson was a respected diplomat and politician, while John Gloag wrote widely acclaimed books on design. These writers came from a social class which feared the consequences of technological progress and all too often displayed their lack of experience by invoking the classic image of the mad inventor.

The visions of the future imagined in these novels were often truly apocalyptic. Shute's *What Happened to the Corbetts* imagined a Britain brought almost to the point of collapse by German air attacks using an advanced navigational system. Neil Bell's *The Gas War of 1940* predicted deaths in the millions in both Europe and the United States. It was by no means the only novel to envision deaths on such a scale, often followed by the complete collapse of civilization. Such visions anticipated the far more realistic warnings that emerged in the age of nuclear stalemate in the 1960s. Shute's *On the Beach* envisioned a world facing extinction from radiation poisoning, a nightmare rendered all the more effective when turned into a motion picture. Imagined worlds where mutants roamed a devastated landscape became regular features in stories by John Wyndham, Walter M. Miller, Jr and a host of lesser writers.[34] By this time, science fiction had emerged as a recognized literary field, and the future war stories merged with those of space flight and other more positive images of technological advance.

The Rise of Science Fiction

The emergence of science fiction as a distinct field of literature is usually associated with the creation of the cheap 'pulp' magazines in interwar America. These published stories of technologically sophisticated futures which – while by no means ignoring the potential dangers – imagined worlds that were richer and more exciting as a result of new opportunities. Here, there were still wars, but there would also be better communications, space travel, perhaps contact with aliens and unlimited power from nuclear or other advanced sources. Life might still have its dangers, but it would be more exciting for almost everyone. Brian Aldiss points out that the focus on America has obscured the contribution of European authors to the same programme. One of the first effective (and widely publicized) motion pictures about space-flight was Fritz Lang's *Frau in Mond (Woman in the Moon)* of 1929, based on a novel by his wife Thea von Harbou.[35]

Lang took advice from the rocket pioneer Hermann Oberth to make his film seem more realistic, a collaboration that pinpoints the difference between the pioneers of science fiction and the pessimistic authors of dystopias. These were writers with a genuine interest in and enthusiasm for new technologies. Hugo Gernsback, who founded *Amazing Stories* in 1926, had gained a technical education before emigrating to the United States. Here, he invented and sold radio sets and published magazines for radio hams and other enthusiasts (his prophetic story 'Ralph 124C 41+' was serialized in his *Modern Electrics* in 1911–12). Gernsback insisted that his authors postulated future technologies that were conceivable extensions of current knowledge. As he wrote in another magazine in June 1929: 'It is the policy of *Science Wonder Stories* to publish only such stories that have their basis in scientific laws as we know them, or in the logical deduction of new laws from what we know.' This policy may have limited the stories' literary merit, but it could also inspire scientists and inventors to try out new possibilities. John W. Campbell, who edited *Astounding Stories* from 1930 and encouraged a higher literary standard, nevertheless agreed with this point, remarking that: 'A concept has to be visualized before it can be realized.' Some magazines had professional scientists as consulting editors. Many scientists of the next generation were turned on to science by reading the pulp magazines.[36]

The enthusiasm these authors displayed for new technologies was in many cases the product of direct involvement with science or technical work. Campbell studied physics at MIT and Duke University. Asimov had a PhD in chemistry, worked at the Naval Air Experimental Station and went on to teach biochemistry at Boston University School of Medicine. Heinlein studied physics and worked with Asimov in the Navy. In Britain, Arthur C. Clarke was a radar instructor in the Royal Air Force and gained a degree in physics from King's College, London, after the war. Those who did not have a science education read voraciously to fill the gap – Frederick Pohl joked that in this way he had acquired as much scientific knowledge as Asimov.[37] Gernsback's hope of creating a field that took the future of science seriously gradually became a reality and readers began to respond critically if they detected obvious flaws.

Whatever their extravagances, the starting point was a genuine appreciation of how scientific knowledge could be applied (see colour plate 5).

This interpretation of the origins of 'hard' science fiction has been disputed by Gary Westfahl, who points out that many of the early stories do not offer realistic projections of the future. The early space operas of writers such as E. E. 'Doc' Smith leapt far beyond the current interest in rockets to postulate interstellar flight driven by totally imaginary devices. Westfahl argues that really hard science fiction only began to be written in the 1950s when writers such as Arthur C. Clarke began to imagine how space flight might actually be developed. This position is hardly tenable – there were some 'hard' stories in the earlier years, including a number about atomic energy. Some of the predictions were so specific they began to cause real concern to the authorities, as when the FBI visited Campbell's office because a story by Cleve Cartmill seemed to reveal what was going on in the Manhattan Project. The pulp magazines were fulfilling a dual function by popularizing both realistic science and mind-blowing extrapolations. Inevitably, there were many misconceptions, including an expectation that space exploration would advance rapidly once it began. Westfahl suggests a number of reasons why even hard science fiction writers have often got their predictions wrong.[38]

The pulp magazines were not the only way these ideas reached the public. Stories about space flight and other novel technologies began to appear in boys' comics early in the century. George Orwell claimed that they really began in the 1930s, but was corrected by Frank Richards, who noted that he himself had written on such themes much earlier.[39] Images of space warfare became more frequent once movie serials such as Flash Gordon popularized the theme in the 1930s. In the 1950s, the British comic *Eagle* had both a space-flight serial, 'Dan Dare, Pilot of the Future' and cut-away images showing the workings of a wide range of technologies, including an occasional futuristic projection.[40] Science fiction authors, whether writing for adults or juveniles, took the possibility of space-flight seriously at a time when everyone outside the ranks of the enthusiasts thought it was moonshine.

The better-crafted stories encouraged by Campbell eventually began to reach a wider audience, and when published as books began to make a wider name for the leading figures. Asimov's career illustrates how the movement developed. He began to write science fiction while still a student and continued after he obtained his job teaching biochemistry. He eventually left his position at the Medical School to write full time both in popular science and science fiction, seeing the latter as a plausible extension of known reality. His early fiction emphasized the possibility of space travel – 'The Martian Way', for instance, predicting exploration of the solar system using atomic rockets. The theme of growing scarcity of resources remained influential – *Caves of Steel* (1954) imagined an Earth where the population lived in vast mega-cities to limit the consumption of resources. It also predicted interstellar travel and continued the theme of intelligent robots introduced in the earlier stories collected as *I, Robot*. The famous 'Three Laws of Robotics' was Asimov's way of challenging the conventional image of the robot as a Frankenstein-like threat. Like Heinlein, who also published a series of stories based on a 'Future History' timeline, Asimov went on to write his best-known work, the *Foundation* trilogy, after reading Gibbon's *Decline and Fall of the Roman Empire*. Here, psychological science in the form of Hari Seldon's psychohistory is added to a galaxy-spanning vision of space travel.[41]

These pre-war authors wrote for a generation of young lower and middle class readers, mostly men and (in America) often recent immigrants who were looking to new inventions as a means of bettering themselves.[42] They read science fiction to imagine a more exciting future, but they also read popular science books and magazines for information about the latest developments in the real world. It was no accident that Gernsback published science fiction stories in his magazines promoting radio and other new inventions, or that the pulp fiction magazines had science advisors and carried articles by technical experts. Popular science and science fiction were both in the business of promoting the idea of technical progress, aimed at the same readership and often written by the same authors.

To create an image of empowerment, this literature praised the efforts of the heroic individual engineer or inventor – even as research was being taken over by governments and big business. The authors, and probably the readers, wanted a better system of wealth distribution organized by experts, possibly following the model proposed by Wells or the Technocracy movement. In Britain, this led radicals such as Bernal and Haldane toward Marxism, while even in America there were a few left-wing writers, notably the Futarian group to which Frederick Pohl belonged.[43] The situation changed during the Cold War. Asimov, for instance, abandoned science fiction for popular science following the launch of Sputnik and the resulting drive to re-educate the American public. His 1964 predictions for 2014 were a natural spin-off from his efforts to show people the effect that science was having on society, and included both the positive and the negative sides of the process.

Popularizing Scientific Progress

The overlap between science fiction and popular science is evident from the output of some of the big names. Like Asimov, Clarke also wrote popular science and made predictions about future developments – in his case famously pointing out the potential use of geosynchronous satellites. There were close links between the fledgling interplanetary societies and the science fiction fan clubs – British enthusiasts such as Clarke were keen to get hold of the American pulps.[44] Popular science had by now emerged as a significant area of publication for both books and magazines, and had been transformed by the innovations in printing that allowed better illustrations. Magazines such as *Scientific American* and *Popular Mechanics* increasingly focused on providing news of the latest developments. Britain had periodicals including *Conquest* (later renamed *Modern Science*) and *Armchair Science*, which had similar aims – although they seem to have been much less successful. In Germany, *Kosmos* and *Koralle* had large circulations, while France had *Science et Monde* and *Science et Vie*. There were also serial works issued in magazine format which accumulated to form substantial surveys and were eventually published in book format. Arthur Mee, better known for the much-loved *Children's Encyclopedia*, edited the *Harmsworth Popular Science*, originally issued in 1911–12 and backed by Britain's leading newspaper proprietor. H. G. Wells and Julian Huxley's *Science of Life* was originally published in serial format.[45]

By its very nature, this material was enthusiastic about research and tried to create an atmosphere in which technical progress was seen as both inevitable and beneficial. The British magazine *Practical Mechanics* was launched in 1933 openly proclaiming this ideology:

But vast as are the strides which science has made in the last twenty years, even more wonderful are the scientific and mechanical surprises which will arrive in the immediate future. The wonder of to-day is the commonplace of tomorrow. This interesting age invites this new publication ... which arrives when all the old orders of thought are changing, creating in the minds of the public a fresh and more intelligent outlook ...[46]

Here was a commitment to the idea of progress, an expectation that the ability of scientists and engineers to change the way we live would continue relentlessly and would produce results that would be both surprising and exciting. The magazines often had up to 10 per cent of their contents devoted to predicting future developments. There were also juvenile periodicals exploiting boys' interests in machinery and these, too, occasionally predicted future developments. *Meccano Magazine* (linked to a popular construction set) sometimes featured futuristic technologies on its covers.[47]

Who could write about science and technology at a level comprehensible and attractive to the ordinary reader? Conventionally, we are told that the increasingly professionalized scientific community of the twentieth century retreated into an ivory tower and regarded writing for the public as a waste of time. Where their nineteenth-century forebears had regarded it as a public duty to spread the word about science, they preferred to leave the job to journalists and writers, who would disseminate the knowledge they produced by translating it into language the general public could understand. But this image is a gross oversimplification, at least for the interwar years. A significant proportion of professional scientists still saw it as their duty to inform the public about science and to engage in debates about how the new knowledge was to be applied. If there was an educational purpose to their writing, the professional community tolerated such activity – as long as it did not preclude research. The only danger was writing for the popular newspapers, where it was hard to get new ideas across without oversimplification or sensationalism. But some eminent scientists, including Haldane and Huxley, did risk writing at this level, and in Haldane's case this was part of an ideological programme driven by his increasingly left-wing convictions.[48]

Although only a few professional scientists achieved this level of public exposure, many more wrote occasionally for the better-quality magazines or for dedicated popular science publications. There was an increasing body of experts with a scientific background who had transferred to the areas of science education and publicity. Walter Kaempffert was the science editor of the *New York Times*, while Gerald Wendt became director of science and education at the 1939 New York World's Fair. Asimov wrote magazine articles on popular science as well as books. Others acquired expertise indirectly – Bertrand Russell supported himself for a time writing popular books on the new physics.

Where publications focused on applied science and engineering, the majority of authors had some technical experience and some were professionally qualified. Some gained technical training in the military during wartime. These experts tended to be politically more conservative than the academic scientists, but were by no means unified in their ideas about how best to promote economic progress. A gulf was opening up between an older generation which saw invention as the job of private individuals and the growing body of technical experts recruited into large-scale government and industrial programmes. The individualists were probably the more active writers both in science fiction and popular science, while those in professional positions were constrained by their employers' interests. There could be considerable tension between the individualists and the professionals. 'Professor' A. M. Low was an inventor and popular science writer who wrote for and later edited *Armchair Science*. He was only 'professor' courtesy of a brief tenure at a military academy during the Great War, and was actually quite suspicious of academic science. He had close contacts with industrialists and thought that the individual inventor still had a role as a source of new ideas that could be taken up commercially.[49]

The journalists who wrote the sensationalized stories so distrusted by the professionals usually had little knowledge of science. Even so, popular magazines often highlighted potentially exciting (or worrying) predictions. The hugely successful British magazine *Tit-Bits* commented on atomic energy in the 1920s and reprinted an article on future technologies by Hugo Gernsback.[50] The *Illustrated London News* also covered atomic energy, along with death rays and movies exploring the possibility of space travel.[51]

These short and sensational features were the despair of working scientists and engineers. But an inexperienced author was unlikely to be able to produce a detailed article, let alone a whole book, on a technical topic. In the early decades of the century, even the most popular newspapers took on aviation correspondents, often drawn from the ranks of the pioneers. The better-quality newspapers were beginning to acquire experienced science editors and efforts were made by the science profession to supply quality copy to the press. In America, Edwin Slossen edited Science Service, a news syndication service set up by Edwin W. Scripps in 1921. Even some untrained journalists became so interested in the field that they built up serious relationships with the scientific community. Many authors of popular books had their manuscripts checked by qualified experts and one cub journalist, Ritchie Calder of the *Daily Herald*, built up contacts that helped him into a career as a science correspondent.

Contact with scientists could generate mixed feelings. Winston Churchill, whose friendship with the physicist Frederick Lindemann began in the 1920s, wrote on biotechnology and atomic power in an article 'Fifty Years Hence' for the *Strand Magazine* in 1931. Acutely aware of the potential for the development of new weapons, he published another popular article under the title 'Shall We All Commit Suicide?' The notoriously gloomy W. R. Inge, Dean of St Paul's, gave a widely reported lecture at the Royal Institution in 1931 predicting the future state of the world in 3000 AD. He hoped for the elimination of all diseases (although he feared the common cold would be hard to defeat). As an enthusiast for the eugenics movement, he predicted a world in which only those with a fitness certificate would be allowed to breed, while the few remaining habitual criminals would be painlessly eliminated. Such hopes were hardly likely to inspire enthusiasm among those with more conventional moral views.[52]

The more positive vision of applied science and technology was often disseminated in the exhibitions held periodically in cities around the world. The British in particular were reminded of the role played by their technical advances in exploiting the resources of their global empire. The British Empire Exhibition held at Wembley in 1938 served this function, although like the New York World's Fair of 1939 its vision of the future would soon be overtaken by war. America, at least, continued to boom in the post-war years, while the British endured years of austerity. The Festival of Britain held in 1951 tried to blend enthusiasm for future developments with nostalgic glances at the world of the past that had been swept away.[53]

Expert Visions of the Future

A number of books were published by scientists and other recognized experts seeking to anticipate what the future would be like as new discoveries were made and exploited. In Britain, the two most imaginative contributions by scientists in the 1920s were both written to promote a materialist alternative to the traditional religious view of human destiny: J. B. S. Haldane's *Daedalus* and J. D. Bernal's *The World, The Flesh and the Devil*. Both became leading figures in the strong left-wing community that emerged in British science between the wars. Their books appeared in Kegan Paul's series 'Today and Tomorrow' along with a number of other works focused on various aspects of contemporary society designed to encourage thought about further developments.

Daedalus was based on a talk given in Cambridge in 1923 and was published the following year.[54] Haldane argued that biology would soon become the key science because it allowed us to transform ourselves in whatever way we chose, either to increase our mental powers or to adapt to changed conditions. The key would be ectogenesis, the artificial production of embryos with deliberately chosen characteristics. Morality itself would be changed to accommodate these new powers. Historian Mark Adams argues that the full range of Haldane's vision was only revealed in his story 'The Last Judgement', published in 1927, in which the Earth becomes uninhabitable as the use of tidal power slows the planet's rotation. Following a vast period of effort, humanity achieves space flight and colonizes Venus, where its newly adapted descendants watch the destruction of their home planet.[55]

Haldane's speculations achieved considerable notoriety. *Daedalus* sold 12,000 copies in its first year and in 1927 Haldane published the first essay in a series on 'The Destiny of Man' in the London *Evening Standard*.[56] His vision of the distant future inspired Stapledon to write *Last and First Men*. But his materialistic vision of humanity taking control of its own destiny horrified more conservative thinkers, being parodied in Huxley's *Brave New World* and serving as one focus for the evil influences combatted in Lewis's *Perelandra* trilogy. Haldane and Lewis clashed in print after Haldane reviewed the novels and protested against Lewis's negative view of science.[57]

Bernal's *The World, The Flesh and the Devil* appeared in 1929, the title proclaiming his materialistic vision of human destiny based on scientific conquest of the material world, the body and the mind. This destiny did not depend on the achievement of a static utopia – the future would be one of constant change. Limitations of earthly resources will lead inevitably to the exploration of outer space, but this move would not wait for the distant future: 'Already ambition is stirring in men to conquer space as they have conquered the air, and this ambition – at first fantastic – as time goes on becomes more and more reinforced by necessity.'[58] Rockets would soon be replaced by more advanced systems and humanity would eventually build permanent colonies in space, exploiting first the asteroids and finally spreading out to occupy the whole universe. All this would be made possible by transforming the human body to adapt to new needs and achieve higher levels of intelligence. Morals would also be transformed: the goal was an almost infinite world of peace and plenty.

Like *Daedalus*, Bernal's prophetic vision gained significant press coverage. The *Daily Herald* reviewed the book with lurid headlines about the transformation of the human body, but also noted the prediction that we would soon move into space, concluding that his book showed that 'Mr. H. G. Wells is but a timid prentice prophet'.[59] The book reinforced the fears of those who saw the new materialism as an overambitious challenge to the traditional view of the human situation. The 'Today and Tomorrow' series had already published *Icarus*, Bertrand Russell's warning against Haldane's assumption that human nature had the capacity to control both body and mind. His *The Scientific Outlook* of 1931 repeated his warnings, but acknowledged the significance of our growing ability to transform the material world.

'Today and Tomorrow' eventually included a number of books dealing with the impact of science and technology, written by authors from a variety of backgrounds. The physicist E. E. Fournier d'Albe acknowledged the social problems of technological development, but predicted that better communications would eventually eliminate wars.[60] American biologist H. S. Jennings warned that the latest developments undermined the credibility of eugenic selective breeding programmes.[61] J. Leslie Mitchell, author of the future-war novel *Gay Hunter*, supported the case for the future of space exploration.[62] Authors with practical experience were not excluded. The inventor H. Stafford Hatfield worried that patent laws were limiting innovation in the area of mechanization.[63] A. M. Low wrote on *Wireless Possibilities*, predicting the mobile telephone and television. Low also wrote two larger-scale futurological works predicting technological developments in a wide range of fields. While praising their value for improving everyday life, he doubted that war would ever be eliminated and accepted that there would be horrors to come. He also thought that telepathy would turn out to have a scientific explanation – and wrote a science fiction novel in which the Martians take over a human inventor via his telepathic projector.[64] This was originally serialized in *Armchair Science*, where Low frequently hailed future possibilities in his monthly column.

Low's were by no means the only large-scale futurological texts produced in the interwar years. In Britain, the war correspondent and conservative politician Sir Philip Gibbs published *The Day after Tomorrow* in 1928, outlining many possible improvements in daily life, but warning about the continuing threat of war. A great deal more publicity was generated by the Earl of Birkenhead's *The World in 2030*, published in 1930. Birkenhead was a former Lord Chancellor, and it seems that his book was ghost-written, borrowing many ideas from Haldane's *Daedalus*. Haldane pointed this out in a review and there was a very public confrontation in the pages of the *Daily*

Express.[65] To be fair, Birkenhead had mentioned his sources and his book was presented very differently from Haldane's. He also had the perception to predict that by 2030 China might become the world's dominant economic power.[66]

In 1933, there was a prediction of future developments for children written by I. O. Evans, who had worked with H. G. Wells.[67] At the same time, Ritchie Calder, a young reporter, was sent by the *Daily Herald* to interview a number of Britain's leading scientists, engineers and medical researchers. He took his assignment very seriously and his articles, which began appearing in February 1932, formed the basis of two books, *The Birth of the Future* and *The Conquest of Suffering*. Although devoted primarily to current research, they contained numerous hints about potential future developments. Calder produced another series predicting developments twenty-five years ahead in 1938.[68] Jonathan Norton Leonard's *Tools of Tomorrow* was equally optimistic.

1936 saw the publication of two substantial futurological works, J. Percy Lockhart-Mummery's *After Us* and John Langdon-Davies's *Short History of the Future*. Lockhart-Mummery was a surgeon, proponent of the theory that cancer is due to mutations in the somatic cells, and a friend of H. G. Wells. He predicted major developments in medicine, the production of artificial food and the need to improve the genetic basis of the human population. Langdon-Davies, who was the author of several popular science books, also focused on future changes to human behaviour and social life and commented on the society envisaged in *Brave New World*.

The focus on biology in many of these predictions reminds us that there was an element of futurology in the eugenics movement that was active both in Europe and America throughout this period. Much eugenic propaganda was aimed at preventing racial degeneration through the excessive breeding of the 'unfit', a movement that reached its horrific climax in the Nazis' effort to purify the Arian race. But Wells, Haldane and many others saw the advantages of encouraging the most able individuals to have more children, with the aim of improving the overall health and intelligence of humanity. The geneticist Herman Muller's *Out of the Night* of 1936 argued for artificial insemination by superior donors. His book was welcomed by many British intellectuals, although it sold poorly in America.[69] Muller and Haldane both saw their left-wing political views as no obstacle to their support for eugenics, provided social inequalities were taken into account when identifying the fit and unfit.

The biomedical angle also featured in claims that individual health and longevity might be artificially enhanced. Muller's efforts to interest Stalin in his proposals failed, but in the early days of the Bolshevik Revolution the hope of improving the race had been enthusiastically supported.[70] The same theme continued in futurological predictions made by writers in Britain and elsewhere and the more extravagant claims occasionally hit the headlines. In 1920, Julian

Huxley wrote for the newspapers about his work on the influence of glands on development. Popular interest focused on Eugen Steinach and Serge Voronoff's efforts to boost sexual potency.[71] For some enthusiasts, medical research literally offered the prospect of immortality, a hope again parodied by Aldous Huxley in his 1939 novel *After Many a Summer*.

The eugenics movement was active in America too, where its warnings about racial degeneration offered a striking alternative to the technological optimism expressed by the early science fiction writers. As in Europe, there were many experts willing to offer forecasts of future threats and opportunities. In 1928, the influential economist Stuart Chase considered the social problems generated by technical developments, worrying about the control and distribution of the increasing mass of material goods and the exhaustion of energy supplies.[72] His concluding 'Balance Sheet' on the impact of science reflected both the benefits and threats. A somewhat more positive view emerged in a collected volume of essays by scientists and technical experts edited by the historian Charles A. Beard. Here, Robert Millikan proposed using solar power to solve the energy problem, while Lee De Forest heralded the potential for radio broadcasting to unite the human race and eliminate war.[73] Beard himself provided a conclusion expressing the hope that with suitable planning, a better world could emerge. An even more positive assessment came from the pen of the Yale chemical engineer C. C. Furnas, whose *The Next Hundred Years* of 1936 predicted the artificial production of food and fuel along with improved domestic appliances and advances in medicine and human genetics. At the time of the New York World's Fair of 1939, Gerald Wendt provided an equally optimistic vision in his *Science for the World of Tomorrow*. In advance of the Fair, the *New York Times* ran a special section of predictions by eminent scientists headed by a H. G. Wells piece on 'The World of Tomorrow'.[74]

By 1940, Europe was already at war, but the more optimistic vision of American science was still being promoted by Walter Kaempffert's *Science Today and Tomorrow*. Kaempffert predicted major advances in medicine, communication and transport, including space travel, but remained skeptical about the practicality of atomic power. By now it was clear that the lone inventor was a thing of the past – the future lay with the big corporations that were increasingly reshaping the world in ways that promoted their own profits. They employed scientists and engineers, but also increasingly designers and public relations experts to promote their visions of the future. The style known as 'streamlined moderne' extended the visual form derived from the latest aircraft and railway trains to almost any everyday object to create a futuristic effect. Designers such as Norman Bell Geddes deliberately extended the range of what was possible by proposing futuristic aircraft, cars and even cities.[75] Their work reached its acme at the World's Fair of 1939, where the Futurama exhibit in the General Motors building allowed visitors to see (via a model) a vision of what the modern city could become.[76]

The Post-War World

After the war, the negative consequences of how the big corporations imposed their vision became increasingly clear after the cities became concrete jungles strangled by the exhaust fumes of automobiles.[77] A new and more critical attitude toward the application of science now emerged. The atomic bomb was the most obvious product of government involvement in big science and public disquiet at the threat of nuclear devastation increased further as the Cold War unfolded. The military became increasingly obsessed with rockets and the hydrogen bomb.

Despite the growing threat of nuclear war, enthusiasts such as Kaempffert continued to make predictions of the benefits that would come if only the problems of population and resources could be solved.[78] Haldane still speculated about the transformation of the human race.[79] Asimov and Clarke supplemented their science fiction with more realistic speculations about what might be achieved. The pro-science lobby remained active, continuing a tradition that had been well established earlier in the century. But its ideological foundations were changing. Haldane's Marxism was no longer acceptable in the era of the Cold War, although some scientists continued to be worried about their profession's involvement with weapons development. Science fiction played a role here, and some scientists were alerted to the dangers of nuclear war by reading the scare stories.[80] Others such as Edward Teller joined whole-heartedly in weapons projects, and the scientific community became increasingly involved with the military-industrial complex. Part of their activity centred on a self-conscious form of futurology in which information-crunching would be used to predict the consequences of various lines of development. The RAND corporation (originally a US Air Force project) and similar institutes allowed figures such as Herman Kahn to articulate the expectations of the military and corporate elites that now dominated the field.[81]

Liberal and humanistic figures were brought into the new futurology. A host of non-military institutions such as the Club of Rome began to look at the wider implications of technological and industrial expansion. These were largely funded by industrial corporations which probably intended their role to be purely cosmetic, but as threats such as environmental pollution were added to the nuclear fears, their publications increasingly began to articulate public concerns. Debate over the real social value of scientific and technical progress thus continued, and some of the concerns articulated in the 1960s and 1970s resembled those of the interwar years.

There were differences of emphasis, especially among those promoting science. There had always been a tension between the individualistic inventors and enthusiasts and the industrial concerns that became ever-more dominant in the real world in which new technologies had to be developed and commercialized. Those corporate interests became even more dominant in the Cold War era. The socialist values that had driven Marxists such as Haldane and some of the early science fiction writers in America were now discredited. The Soviets had tarnished the Wellsian image of a rationally organized state and given comfort to the moralists who distrusted science as a path to totalitarianism. The military-industrial complex that had emerged during the war now controlled the most powerful efforts to promote scientific progress as the key to a better future, forcing the science fiction community to move further across into the anti-progress camp. Cold-War-era science fiction began to articulate a more nuanced image of scientific progress. It still expressed the hope that we might go out into space, but it also expanded its mission to warn about the potential dangers of an unrestricted rush to conquer the natural world. Concerns on this front had been articulated earlier in the century, but now they became a major theme.

3   How We'll Live

The most obvious place to start is the most prosaic: how did people imagine their daily lives would be changed by new developments in science and technology? Even here it was by no means easy to work out what people thought would really make a difference. Enthusiastic inventors proclaimed the benefits of applied science, but in the future depicted by many literary figures few ordinary people would see those benefits. In 1907, Jack London's *The Iron Heel* predicted an America in which the middle class had been eliminated and the workers enslaved by an increasingly repressive plutocracy. Were we all to become slaves of the machine to serve the purposes of the rich, or would the benefits of mechanization spread to all? By the 1920s, Wells no longer thought that the workers would become a permanent underclass. Attention now focused on how the system of distribution could be reformed to make the flood of goods produced by mechanization available to all. There were setbacks, of course, and in the worst years of the Depression many found it hard to believe that their lives would improve. But there were schemes to reform the economic system, and the example of Soviet Russia showed that this was possible even if one didn't approve of full-blown Communism. In this environment, the enthusiasts could make more of the running.

The promoters of applied science assumed that eventually the mechanization of industry would give everyone more leisure time (even if the work involved only boring routines). The dirty and unhealthy factories of the first Industrial Revolution would be replaced with new structures that were better to work in as well as more efficient. It was also predicted that there would be more gadgets to make life easier in the home. Popular science magazines routinely printed lists of inventions designed to make life easier or more enjoyable, most of which turned out to be useless and sank into oblivion. Inventors and promoters sang the praises of their products, but they were competing against one another and success for one would block the chances of others. Some promotional material was purely superficial, as with the enthusiasm for streamlined designs applied to objects that didn't move. Even when they were on to a good thing, technical experts and marketing agents were unlikely to foresee the wider consequences of a revolutionary technology. Yet in a few cases – radio broadcasting is a good example – almost everyone could see that their lives could be significantly transformed by what was promised. What couldn't be predicted were the deeper consequences for society, which is why some commentators worried about things that the majority greeted with enthusiasm.

The image of the 'ideal home' transformed by electricity, radio and television was the most obvious vehicle by which new technologies were promoted to the public. Forward-looking industries also looked for modernized offices and factories to symbolize their efficiency. The actual buildings within which people would live and work became the subject of technical developments, some of which would have a real impact on the lives of those who lived in them. New construction techniques pioneered in America created the skyscraper city, expanded in the imagination of figures such as Wells and Asimov into the giant mega-cities of the future in which everyone lived in apartments and ate in communal refectories. Industrial activities would be confined to the lower reaches where noxious by-products would cause less offence to those living above.

New transport systems worked in the opposite direction, however, opening up the city to suburban sprawl or, more imaginatively, to a system of small and entirely new garden cities. Factories would be dispersed, and would be made to blend into the environment. Management of the traffic flow when most individuals had private motor cars soon became an all-too-real problem for governments and planners. Huge excitement centred on the hope that long-distance travel would be opened up by aviation. Some hoped that private aircraft would be available for all, but seldom thought about how air-traffic control would be managed.

Beyond these promises for new machines and living spaces lay the growing hope that medical science would eliminate disease and extend our vigour or even our lifespan. Here, again, the wider social consequences were hard to predict. Only the bolder thinkers ventured so far into the future. For ordinary people who got their ideas and information from newspapers, magazines and eventually the radio, it was the down-to-earth promises of inventors and manufacturers which offered the most visible window into the lives they might hope to lead in the near future. At the domestic level, regular Ideal Home exhibitions gave manufacturers a platform from which to advertise their latest products by painting a picture of the better world to come.

Drudgery or Unlimited Leisure?

The interplay between the mundane and the more imaginative forms of futurology was most apparent in the uncertainties surrounding the balance between work and play. Along with London's *The Iron Heel* and the early stories of H. G. Wells, Fritz Lang's movie *Metropolis* presented images of a world in which mechanization became a curse to the working class – they were reduced to little more than slaves toiling over the machines to benefit the rich who controlled the system. London's book was still being reprinted in the interwar years with a preface by Anatole France claiming that its predictions were starting to come true.[1] America was certainly seen as the model for the new industrialized society, not always with enthusiasm. French thinkers in particular vacillated between excitement and fear that Americanization would destroy the national culture.[2]

While the enthusiasts for the ideal home took it for granted that the benefits of technology would spread to all levels of society, radical political thinkers worried that the capitalist system would resist progress and deny the workers any hope of a better life. This concern was hardly surprising given the example of the previous century, but it turned out to be unfounded. No political figure in a democracy could openly promote a deliberate limitation of the benefits of science to the upper classes, and dictators of both the right and the left derived their appeal from the claim that their rule would benefit the majority. In the end, improved ways of running the household did gradually spread more widely, despite the upsets generated by an unregulated capitalist economy.

The weakness of the most pessimistic scenario was exposed by industrialists such as Henry Ford, who showed that mechanization and a systematic organization of the workplace vastly increased productivity, which was pointless unless the workers simultaneously gained the ability to buy the goods flooding out. The rich didn't need mass-produced goods – they were the only ones still able to afford things made by real craftsmen (which in turn made it possible for the artist and skilled artisan to survive).[3] By the 1930s, there were claims that applied science was now so successful that shortages of food and goods were a thing of the past. The Depression showed that there were still major problems in how the goods were distributed, but the actual supply was potentially able to provide comfort for all. Left-wing commentators pointed to the destruction of surplus food as evidence of the wastefulness of capitalism, but others saw it as a sign that shortages were no longer a problem.

Work on the assembly line might be boring and regimented, as depicted by Charlie Chaplin in *Modern Times* (1936), but there would be less of it and hence more leisure time, as well as better wages to spend on the motor cars and other products of industry. In the 'Today and Tomorrow' series, H. Stafford Hatfield presented the negative side of automation, while Garet Garrett offered a more positive approach based on the increase in productivity. The Earl of Birkenhead predicted a ten-hour working week by 2030, while John Langdon-Davies thought three hours' work per day would suffice by 1960. C. C. Furnas called for less fear of machines displacing human workers, arguing that mechanization would cut the working day in half as well as providing a better standard of living. In a 1931 editorial for *Astounding Stories*, Hugo

Gernsback rejected the popular assumption that mechanization was responsible for the Depression.[4]

Concerns about the mechanization of industry were often identified with science fiction's image of the robot as a threatening mechanical monster. There certainly were efforts to create 'mechanical men'. In 1934, the British magazine *Practical Mechanics* featured an image of a robot on its cover and five years later reported on Elektro, an 'almost human' robot exhibited by Westinghouse at the New York World's Fair. In *Our Wonderful World of Tomorrow* of 1934, A. M. Low included a chapter on 'The Robot Age', arguing that industrial robots would not be the humanoid machines imagined by the fiction writers.[5] They would be designed to do specific tasks automatically in order to boost production. Low mentioned the possibility of robots taking over domestic tasks, but only briefly. The popular science literature of the interwar years contained relatively few descriptions of robots capable of replacing the housewife. Attention focused instead on having the chores eliminated by built-in facilities such as automatic vacuum cleaners, while cooking would be replaced by pre-prepared meals.

The prospect of autonomous robots taking over the running of the household became more plausible in the post-war era and was most clearly articulated in Isaac Asimov's series of robot stories. He introduced the 'Three Laws of Robotics' in his collection *I, Robot* in 1950 explicitly to challenge public fears engendered by earlier science fiction writers. The stories would help to make the public feel more comfortable with the prospect of sharing their lives with thinking machines. Robots would certainly take over all domestic chores, and in 'Satisfaction Guaranteed' Asimov introduced one that learns enough to become a fashion and lifestyle advisor to a timid housewife (unfortunately she also falls in love with it). In 1967, Herman Kahn and Anthony J. Wiener suggested that domestic robots would be commonplace by 2000.[6]

Writing of the 'robotization' of industry in 1929, Stuart Chase noted that only 12 per cent of the workforce was actually employed on such repetitive tasks anyway. Ralph E. Flanders argued that both work and everyday life would become more pleasant as better design eliminated noise and pollution. Low, whose main area of expertise was sound and noise reduction, predicted that improved mechanisms would soon eliminate unwanted noise in both industry and the home, a theme also taken up by Richie Calder. The designer Walter Dorwin Teague argued that machines gave us the power to create order and purpose in industry and would restore the beauty in products lost during the first Industrial Revolution. Lewis Mumford saw a new wave of industrialization coming that would be better adapted to human needs.[7] There was hope that industrial psychology might make even work on the production line less soul-destroying. Optimists saw this as beneficial to the individual, although more cynical eyes saw the threat of psychological manipulation to create a pliable workforce.

The ordinary person might not have much choice in how the new society would be organized (see Chapter 11). Ostensibly, democracy was the dominant ideology of most Western countries, but the interwar years saw increasing threats to its influence. The totalitarian regimes that came to power in Russia, Germany and Italy provided different models of how an intrusive state might justify its actions. The plutocracy depicted in London's vision of the American future did indeed expand its power and influence, although it gradually came to appreciate that it was counterproductive to reduce the working class to misery. Many intellectuals followed Wells in calling for government by experts on the grounds that democracy was an ineffectual check on the shortsightedness of the capitalist system. Yet in their various ways most of these rival systems claimed to offer benefits for the masses – except, of course, for those elements deemed undesirable. Most workers would be offered a better life, even if what they gained would be limited when compared with the lives of their rulers. Birkenhead thought people would gladly accept the rule of experts provided major decisions were put to a referendum.[8] The Depression exposed the weaknesses of the existing system, but many escaped the worst evils of unemployment and for much of the period it was possible to hope for a better life.

What was on offer certainly included a more comfortable life in the home, especially through the use of electricity. Whether that home would be a conventional house, some futuristic construct or an apartment in a huge metropolis was up for debate. Better entertainment would be provided, with much attention focused on the prospect of television. Outside the home there would be more freedom of movement, with motor cars and perhaps even personal aircraft becoming available. All of this presupposed more leisure time, and there was a good deal of debate among intellectuals and cultural pundits on how ordinary people would respond to this opportunity. Some hoped that they would seek informal education and active involvement in the arts. The radio pioneer Lee De Forest called for broadcasting to be used to spread culture and in Britain the BBC explicitly set out to do this. Others thought that the state would insist on better education for all. Birkenhead predicted universal education to university level, while Langdon-Davies thought there should be compulsory education to age 21.[9] Many commentators also insisted that it would be necessary to encourage sports to keep people active.

Where commercial interests reigned, broadcasting adopted a less highminded approach. Radio and later television responded primarily to what people wanted, especially popular music and comedy shows. Many intellectuals feared the emergence of a passive society of consumers who would allow the entertainment industry to shape their world in ways beneficial to the state or big business (see Chapter 5). The lower classes in Huxley's *Brave New World* had plenty of 'free' time, but were conditioned to spend it in mindless and resource-consuming games.

Electrifying the Ideal Home

By the 1920s, the lives of ordinary people were already being improved by the technologies introduced since the turn of the century. In 1927, there were 26.7 million households in the United States, of which 11 million had a phonograph, 10 million a motor car and 17.5 million a telephone.[10] Europe lagged behind, but its people were encouraged to hope that these amenities would soon become available. Invention was the key to a better life, and every issue of popular science magazines such as *Scientific American* and the British *Conquest* listed new gadgets that were designed to make things easier. Many were of little significance – in its first issue, *Conquest*'s feature 'Novel Patents and New Ideas' included a device for cleaning forks.[11] Even so, the constant reporting of new developments created an air of anticipation.

The ideal home was almost invariably hailed in the press as 'the home of tomorrow', something that would become available in the very near future. A key feature of the new domestic technologies would be the application of electricity. In the late nineteenth century, electricity had already emerged onto the public consciousness as a source of power that might transform the way in which we live. J. J. Astor's *A Journey in Other Worlds* of 1894 had depicted life in the year 2000, when transportation and other aspects of life had benefitted from the applications of electricity. Inventors such as Thomas Edison and Nikola Tesla gained immense reputations as the wizards who would direct the change. The creation of power grids expanded over the early decades of the century in both America and Europe, allowing an ever-widening section of the population to participate in the expectations.[12] Since most work around the house was done by women, much of the promotional literature was aimed at them. As Thomas Edison wrote in *Good Housekeeping* in October 1912:

The housewife of the future will be neither a slave to the servants nor herself a drudge. She will give less attention to the home, because the home will need less; she will be rather a domestic engineer than a domestic laborer, with the greatest of all handmaids, electricity, at her service. This and other mechanical forces will so revolutionize the woman's world that a large part of the aggregate of women's energy will be conserved for use in broader, more constructive fields.[13]

The horrors of the open fire and the solid fuel cooking range would be a thing of the past. Magazines and books were soon proclaiming the imminent fulfilment of these promises. In January 1920, *Conquest* hailed the introduction of electricity for cooking and heating and noted that an electric dishwasher already available in America would soon come to Britain. A dishwasher was also included in a French survey by André

Christofleu predicting family life in the near future. In the early 1930s, A. M. Low was still proclaiming the benefits that would be gained from electric living, with an image showing a well-to-do couple breakfasting with electric coffee pot, toaster etc. (Figure 3.1). He also predicted built-in vacuum-cleaning systems that would also sterilize the air.[14]

A particular source of concern for the advocates of applied science was the lighting of both home and public spaces. The electric filament lamp produced light only as a by-product of much wasted heat. What was needed was a source of light that was both more natural in quality and more economical, perhaps derived from techniques mimicking the bioluminescence of some living organisms. This desideratum was anticipated in the *Harmsworth Popular Science* in 1912 and was a regular feature of post-war visions of the future by writers such as Low and Gernsback. In 1928, *Popular Mechanics* reported an exhibition depicting the home of the year 2000 lit by a new form of glass, which would not only let in more sunlight, but would actually produce light on cloudy days. By 1939, fluorescent light was presented as at least a partial solution to the problem and was hailed by Low in a 1951 book on electronics.[15]

Low also noted another new invention that would indeed transform our lives – the electronic oven, now better-known as the microwave. This also featured in a short film introduced by Low entitled 'Science is Golden', in which the (all too conventional) housewife inserts pre-prepared food to be heated for her husband's dinner. Even before the war, designers such as Le Corbusier had enthused over the 'new kitchen' and more rationally designed furniture. By 1962, Arthur C. Clarke moved speculation onto a new level by predicting a device to reproduce food from stored information defining its chemical and physical structure, later popularized as the 'replicator' of the *Star Trek* television series.[16]

The ability of electric power to coordinate mechanisms was also celebrated. Christofleu hailed the automatic valet de chambre in which the alarm clock would close the window, light the fire and boil the kettle. Low's cartoon poked fun at the idea by suggesting an alarm clock rigged to drag the bedclothes off a reluctant riser. In 1939, *Popular Mechanics* predicted an all-electric home in which all the appliances would be remotely controlled. It also foresaw electro-static dust removal and a similar process for cleaning clothes. In 1942, it promised regular sterilization of the home by ultra-violet light. By 1967, the various facilities of the all-electric home of the near future were to be coordinated by a computer in the basement, programmed in accordance with the owners' regular habits.[17]

Fig. 3.1. Cartoons showing the pros and cons of the all-electric household. The convenience of the breakfast table is offset by the alarm clock that strips the bedclothes off if you don't get up. From A. M. Low, 'Women must Invent', *Armchair Science*, March 1932, pp. 674–6.

A Synthetic World

Washable furniture was an extension of another trend that fascinated the enthusiasts for applied science: the production of new materials, especially synthetics. Home equipment, clothes and even food would all be transformed by the power of chemistry. Low devoted a chapter of his *Wonderful World of the Future* to the prospect of new synthetic materials. He also proclaimed that 'Artificial Life is Best' and professed a preference for artificial flowers over the real thing.[18] Plastics became the most visible symbol of this potential, but the trend had already begun long before they became widely available in the 1940s. At the height of the promotional frenzy, plastics were endowed with a trans-formative power in their own right – they were revolutionary new materials, not mere substitutes. They allowed the creation of new designs both for existing household equipment and for entirely new gadgets. Eventually, the promises turned out to have been over-stated. Plastics became a by-word for cheap and shoddy products whose streamlined designs were little more than a marketing ploy. The exaggerated claims for the artificial production of foodstuffs also failed to be realized.

The expectation that applied science would transform the objects we con-sume was a continuation of earlier trends. Substitute materials were increas-ingly employed from the beginning of the century, especially in the construction industry, but were of limited value. Writing for the 'Today and Tomorrow' series in 1928, H. J. Birnstingl dismissed many of the substitutes as 'compressed sawdust or congealed offal – conglomerated dung or coagulated sewage', offered for sale by charlatans encouraged by governments desperate to alleviate housing shortages.[19] Bakelite, which became available in the 1920s, was the first plastic to achieve real success and was used for many household items. Plastics were hailed as cleaner, cheaper and more versatile than traditional materials. By the late 1930s, streamlined designs were being used to give household items a modern or even futuristic appearance, made possible by the use first of aluminium and then by the ever-expanding range of plastics. Some uses were quite imaginative: in 1928, *Popular Mechanics* predicted inflatable chairs and beds, and in 1942, washable furniture. Plastics had now become the dream material that would revolutionize life. It was widely claimed that we were moving into an 'age of plastics'. Kaempffert saw the production of synthetics as a chemical revolution and wrote of the 'Chemical Revolt against Nature'. The advice given to Dustin Hoffman's character in *The Graduate* as late as 1968 was: 'Just one word . . . plastic . . . there's a great future in plastics.'[20]

The design potentials of the new materials were celebrated by Paul T. Frankl in 1930.[21] Plastics were the ideal material to give objects a smooth, rounded appearance, although streamlining has no real purpose in a static object.

The new designs were intended to give consumers a sense that they were moving into a new world. They were a deliberate ploy used by designers and marketing departments to get the public buying again after the Depression. The role played by the advertising industry in promoting the latest materials and gadgets was noted by John Langdon-Davies in 1936 – writing of everyday life in 1960, he ridiculed the ways in which consumers' desires were manipulated by suggesting how silly it would be if we were persuaded to buy different brands of drinking water![22] There were many useful applications of the new synthetics, but doubts emerged as early as 1944 when Walt Disney released a Donald Duck cartoon in which the invention of a new plastic turns out to be a disaster.[23] By the 1970s, plastics were no longer seen as the key to a better lifestyle. Anyone with taste (and enough money) now thought traditional materials were better.

A particular focus for the application of new materials was clothing. Here, the move to synthetics was an extension of wider transformations promised for everyday life. Those who saw science as a positive agent of change thought it inevitable that rational attitudes would soon be applied to clothing styles, sweeping away the over-elaborate and inconvenient modes of dress inherited from the past. Birkenhead called for a move toward less restrictive clothing, while Low stressed the need for more hygienic clothes and imagined a world in which everyone – men and women – wore electrically-heated boiler suits. Writing for children, I. O. Evans thought future clothes would be more sensible and more colourful (even for men). He emphasized that this would become possible because new synthetic fabrics would be produced, outdoing the artificial silk that was already on the market.[24] Plans for new fabrics had begun long before the age of plastics. In 1913, *Popular Mechanics* told its readers that the chemist William Perkins had developed a fireproof cotton that would even work better for women's dresses. In 1929, it hailed the prospect of clothes made from asbestos, while a few years later, *Armchair Science* predicted they might be made from fibreglass.[25]

In the end, though, it was plastics that led the way, with nylon making the first major inroads into the clothing industry in the 1940s. The craze for nylon stockings has become legendary, although men eventually decided they didn't like shirts made of the stuff.[26] The drive for synthetics was parodied in the 1951 Ealing comedy *The Man in the White Suit*, in which Alec Guinness plays a chemist who invents a material that repels dirt and apparently never wears out. The mill-owners and unions are outraged because no one would ever need new clothes – but the material self-destructs after a few days. The weekly wash was in any case becoming less of a chore by the late 1940s thanks to the first synthetic detergent, *Tide*. More in the realm of science fiction, the use of ultrasound to clean clothes was being anticipated by 1948.[27]

What We'll Eat

The other area of daily life which promised to be transformed by technology was the food supply. Here, again, synthetics were promoted as the most radical innovation, but their impact was actually less than that of other techniques transforming the ways in which natural foods were preserved and distributed. Scientific agriculture had already transformed food production, allowing some to claim there was little fear of shortages in the future unless the population expanded dramatically (see Chapters 10 and 11). This point was noted by J. P. Lockhart-Mummery in 1936 – he did worry about population expansion, but others were concerned that the population was declining as better-off people had fewer children. Langdon-Davies thought food and other necessities would become so abundant they would be distributed free – they would be as cheap as air and we will 'eat as we breathe'.[28]

As the threat of shortages seemed to diminish, attention shifted to the preservation and distribution of food. Ritchie Calder hailed the long-established technique of canning as the best way of preserving food and guaranteeing a nutritionally balanced diet for all. He admitted that freezing was becoming more practical as a means of preserving fish and meat, and predicted that in thirty years (from 1934) most food would be supplied in a preserved form. In 1937, *Popular Mechanics* predicted that most food would soon be supplied frozen and ten years later it anticipated what soon became the TV dinner. Calder called for better methods of preserving fruit, but suggested that in any case it would soon be transported rapidly around the world by air. Fruit from South Africa would reach London in a day by 'express goods-planes pulling a train of glider trucks'.[29]

Despite this apparent plenty, technophiles became obsessed by the prospect of producing artificial food. This would not only be cheaper, easier to distribute and possibly more healthful, but would also remove the threat of starvation once and for all. The artificial production of sugar from cellulose attracted much attention in the 1930s and led to expectations that other foodstuffs could be synthesized. Haldane and Birkenhead noted this prospect and foresaw a time when agriculture would become just a rich man's hobby.[30] In 1928, *Popular Mechanics* reported that Professor James Norris of the American Chemical Society had predicted the synthesis of foodstuffs, citing the example of German efforts during the Great War. Lockhart-Mummery waxed enthusiastic about the artificial production of food from cellulose, but conceded that huge forests would be needed to supply the wood. Proteins would be more difficult to synthesize, but he thought science would eventually solve the problem. *Armchair Science* reported the synthesis of sugar from wood and even from chalk. Low hailed the hydrogenation of coal as a means of producing not just oil, but also synthetic foodstuffs. In 1940, the addition of chemically modified grass as a food supplement was proposed in *Popular Mechanics*.[31]

So ambitious were the suggestions for artificial food that serious commentators felt it necessary to warn against press reports that soon we would take all our nourishment in the form of pills. Furnas and Lockhart-Mummery both dismissed such reports as nonsense, insisting that the human digestive system needed bulk to function properly. They pointed out that we want flavour and variety, but insisted that the chemists would be able to supply their synthetic foods with any desired flavour. Evans also pointed out the need for bulk to his younger readers, although he admitted that food pills would be useful to travellers in remote regions. Two volumes in the 'Today and Tomorrow' series dismissed the press speculation as nonsense, the one devoted to food ridiculing the prediction of a world of toothless people dining on 'a tabloid of artificial protein, chemical lemonade and an injection of strychnine'. The book ended with a fictional report from a future devastated by rampant vegetarians. From the perspective of the chemist, T. W. Jones also dismissed the 'tabloid theory' and argued that science would merely help increase the production of normal foods.[32]

In the end, there were indeed limits to how far the synthesis of food could be taken and the world continues to get most of its supplies from natural sources. By the 1950s, the fear of shortages re-emerged even though agricultural science continued to improve yields. The global population was expanding rather than contracting, mainly because the benefits of an improved lifestyle had not disseminated beyond the advanced nations. There were now calls for scientific agriculture to be supplemented by new techniques such as hydroponic farming, the cultivation of algae and bio-engineering. Asimov's 1964 vision of the world fifty years ahead expressed these growing concerns (his novel *The Caves of Steel* had already predicted Earth's population huddling into massive, crowded cites under the pressure of diminished resources).[33] In the developed world, the chemical manipulation of food products proceeded apace, ostensibly to improve quality and preservation. But the ever-increasing number of artificial chemicals employed provoked growing resentment and regulation in the later decades of the century, a story that lies outside the range of the present study. If the age of artificial food did not emerge as the enthusiasts imagined, the modification of traditional foodstuffs made it increasingly difficult to identify what is really 'natural'.

Women's Work?

There is little doubt that home life did become more comfortable for many, thanks to innovations such as electrification. Coupled with the spread of broadcast entertainment (discussed in Chapter 5), the middle and better-off working class family now lived an easier and richer life. Much of the advertising for the new products was specifically targeted at women. The traditional notion of the 'housewife' encapsulated the assumption that most domestic chores were hers, which is why Edison promised in *Good Housekeeping* that women especially would benefit from electrification. The popular science literature of the early twentieth century was replete with similar promises that women's household chores would become easier, giving them extra time for leisure and to improve themselves. The most forward-looking anticipated a complete emancipation allowing women to participate equally in the workforce and to have careers in business or in science and invention.

One of the most vociferous proponents of the view that more technology would liberate women so they could become the equal of men was A. M. Low. He foresaw a world in which '[w]hat is generally called "housework" will be eliminated by a hundred different devices'. Women no longer served as housewives and would demand equal opportunities in education and careers – Low thought there might even be a female prime minister. They would abandon their frivolous concern with fashion and might even take to wearing trousers. He also insisted that women must become inventors if they wanted to ensure that new technologies would fit their own requirements.[34] Evans told his young readers that the new technologies would free their mothers from housework.[35]

As Low realized, it was not just the reduction in household duties that would make the transition possible. New attitudes toward sex and child-rearing would also promote liberation from traditional constraints. Birkenhead confessed that he was no supporter of feminism, but conceded that if women could free themselves from constant childbirth, they could demand to work alongside men outside the home.[36] Among the most radical prophets was Langdon-Davies, who predicted the complete separation of sex and marriage, the disappearance of the family unit by 1975 and the transfer of child-rearing to the state. He was well aware that these claims chimed with the dystopian predictions in Huxley's *Brave New World* – 'all the restless imagination and thoughtful torment of the frustrated scientist' – but insisted that these developments were the natural consequence of applying rational thinking to social questions and it was short-sighted just to dismiss them as nasty.[37]

Some who worried about long-range transformations in society saw the possibility that liberated women might come to play the dominant role.[38] The artificial generation of children did not happen, of course (although we seem to get ever closer to it nowadays), but Langdon-Davies was by no means the only commentator who saw that changes in how men and women interact would be reflected in new attitudes to child-rearing. Most of the advertising for new household products presupposed that they would be used in the conventional family home, either a house or at least a city apartment. But there were changes at work that threatened the way in which people would be housed, and some of these impacted directly on the possibilities for social transformation. Perhaps we should all be encouraged to live cheek by jowl in vast cities. In these circumstances, it would be natural for some household functions, most obviously cooking and eating, and the rearing of children, to be transferred from the family unit to the community. As the next chapters will show, these speculations led to a real anticipation of life in a world where a much more limited range of activities went on in the home.

Here was a prospect that would really liberate the housewife – most of the chores would either be done automatically or had been dispensed with altogether in the drive toward communal living. Of course, it didn't work out like that, and even at the time there were critics who pointed out how women's expectations were being manipulated by advertisers desperate to promote their products by promising them an easier life. Big business did not want to encourage the sort of communal living associated with the Soviets (or with the *Brave New World*) because that would limit the range of goods that could be sold. So women were endlessly targeted with advertising claiming new gadgets were 'labour-saving'. In 1928, H. J. Birnstingl wondered who had coined this misleading term and criticized the exhibitions, periodicals and commercial travellers seeking to convince women that buying the products would make their lives easier. The women themselves were also to blame: 'the credulity of women knew no bounds', he proclaimed.[39] He admitted that some innovations were beneficial, although he ended his little book with a spoof account of housing in 1987 which included a move toward deliberately antique styles.

Ruth Schwartz Cowan has shown that women's work in the home did not, in fact, diminish and for a generation at least the housewife still slaved to keep her family comfortable.[40] As prosperity grew, domestic servants became less affordable by the middle classes, so the lady of the house increasingly had to do things which in Victorian times would have been delegated to housemaids. Far from reducing her workload, labour-saving gadgets merely allowed her to keep up with the extra demands on her schedule. She may have had fewer children, but looking after them became more demanding as out-of-school activities flourished. The poor still had larger families and could not afford the new technologies, or had to settle for inferior brands that didn't have the same benefits. Probably the best outcome was for the better-off working class, where the women stayed at home but could make use of at least some of the new developments. Only in the later twentieth century did the move to give women better access to careers outside the home begin to take off (wartime excepted).

4   Where We'll Live

Predictions of new household gadgets to make life easier were only part of the promise that applied science had to offer. Work would also become less onerous and there were expectations that the physical form of both homes and factories might be transformed. People had always lived in houses of various levels of comfort and these structures might be improved to make living in them more attractive. The same was all-too obviously true of the older factories and industrial plants. New materials made new designs possible and might even allow all or parts of the buildings to be moved around. New dispositions for living also became possible. As construction techniques improved, great skyscrapers began to transform city centres, especially in America. The prospect that we might all live in apartments within a vast megalopolis emerged, and with it the possibility that many activities would be transferred from the home to a communal forum. At almost the same time, however, new methods of transportation began to undermine the trend toward more centralization. Once it was possible to commute significant distances in an hour or so, there seemed no reason why the distinction between city and countryside might disappear as people moved out into suburbs, garden cities or more dispersed tower blocks.

What was happening can be seen as a classic example of the interplay between technical innovation and wider economic, social and cultural forces. Rival technologies made different futures possible, allowing commentators from various backgrounds to make conflicting predictions about what would actually happen. Steel made possible the dense, vertical city of skyscrapers, but other new materials were best adapted to improving conventional housing. H. G. Wells foresaw the concentration of the population into great metropolitan centres in his *The Sleeper Awakes*, but soon realized that faster transportation might make this unnecessary. At the opposite extreme, Rudyard Kipling's stories of the Aerial Board of Control were set in a world where personal aircraft allowed almost everyone to live in isolation.

Enthusiasts disagreed on how to exploit the possibilities of the new technologies. Wells became a spokesman for those who argued that experts should be called in to plan social developments along rational lines. Various schemes were proposed to achieve a dispersed population, each reflecting a different social ambition or ideology. The flight of the well-to-do into the suburbs was all too often unplanned, but those who exploited the situation for private profit were challenged by environmentalists and town planners, who pointed to the benefits of well-structured communities or brand new garden cities. Some high-tech industries used new building designs to overcome the image of the 'dark satanic mills' left over from the nineteenth century. The image of dispersed city-blocks promoted by Norman Bel Geddes's 'Futurama' display at the 1939 New York World's Fair was a kind of hybrid fusing the benefits of better transport and a more imaginative use of construction techniques.

The social consequences of the alternative scenarios were by no means easy to work out. Wells was not the only writer to see the megalopolis as a place where a downtrodden working-class lived in poverty and slaved in underground factories. Yet city-dwellers had been living in apartment blocks for centuries (millennia if we include the Romans) and there was no reason to assume that the majority would be reduced to the level Wells predicted. Nor did it seem reasonable to imagine all heavy industry buried in caverns beneath the levels in which people lived. Optimists saw the city as a place of plenty and opportunity for all and even celebrated the creative power of a crowded humanity. Huxley's brave new world was urbanized, but with the population conditioned to enjoy whatever pleasures were offered to their particular genetically predetermined social class. At the opposite extreme, Kipling's aviation-savvy citizens were aggressive individualists, while the promoters of innovative designs and garden cities lamented the preference of so many for backward-looking suburban houses. There was much antagonism between the advocates of aggressive futurism and those who wanted a more subdued modernism.

Because the interplay between technology and social forces was so complex, this was an area where predictions were frequently wide of the mark, and were exposed as such within a few years. Wells realized almost immediately that his prediction of giant mega-cities might be negated by better transportation. Hopes for marvellous new designs for prefabricated houses were shown to be wildly optimistic by the 1940s. In the end, the situation was so complex that almost all of the predictions turned out to be partly valid, provided one waited long enough for fulfilment. Cities have indeed grown much larger, surrounded by increasing urban sprawl, endless bland suburbs, and a few isolated planned communities.

Building for the Future

The least imaginative view of future living envisioned the majority of families still in individual houses, although they would no longer be situated

immediately alongside the centres of industry. Most popular science literature, especially in Britain and America, was aimed at the individual house-
holder. This was partly a reflection of cultural backgrounds – the British
city-dweller had never favoured the continental love of apartments and in
America the pace of urbanization had only recently begun to pick up. Strong
corporate interests saw the self-contained family unit as the best source of
sales possibilities.

The house itself offered plenty of opportunities for technological innovation,
especially when demand for new living spaces was intense (most obviously in
the aftermath of the two world wars). The emergence of new industries also
opened up the hope that the workplace of the future would be clean and
efficient. New materials were developed, allowing more imaginative and
more practical designs for both homes and factories. Innovative architects
devised modernist styles to challenge the desire of many ordinary people for
a home reminiscent of the past. Industrialists positively welcomed new designs
that would symbolize their forward-looking ventures. Prefabrication became
possible and it was widely predicted that soon houses and even factories would
be erected in a few hours and could be moved at the drop of a hat. Some thought
we would soon all be living in mobile homes.

Whatever the lure of the skyscraper, architects and town planners realized
that there was still considerable demand for the conventional house with its
own garden. New materials opened up the possibility of transforming the
traditional family home. Prefabrication offered homes that were cheaper to
build, but also an improvement on the dismal environment in which most
people had lived. Glass and plastic would allow a much lighter form of
construction, making the house more pleasant to live in and the whole structure
less permanent and restricting. The interior could be reshaped at will, or even
the whole house transported from one site to another. The interior would be air-
conditioned, of course. Mobility became a goal for many designers in the
interwar years. Yet here again rival technologies offered different ways
forward – if the interior was to be cut off from the outside world, some planners
envisioned houses moving underground to save energy on heating and to
protect from storms (or, later on, nuclear attack).

There was certainly a demand for new housing. In America, the population
was expanding rapidly, while the aftermath of two wars led to huge demands in
Europe too. Soldiers returning to Britain after the Great War were promised
a 'land fit for heroes', but both jobs and decent houses were in short supply.
To many designers, it seemed clear that the best way forward was to abandon
traditional bricks and mortar in favour of prefabricated housing that could be
erected rapidly on more or less any site. New materials were developed to speed
up the process, although the projected advantages often turned out to be
illusory. The damning comments quoted from H. J. Birnstingl's contribution

to the 'Today and Tomorrow' series in the previous chapter were aimed in part at these substitute building materials. For him, the new opportunities benefitted the building industry, not the occupants, and the whole process had turned into a 'jerry-builders' paradise'.[1] Birnstingl appreciated the possibilities for improvement, but thought they were being thrown away in the name of expediency. America offered a bad example, although he praised German efforts to combine industrialism with beauty. In part, the problem was created by the public themselves, since many actually preferred old-fashioned designs which created an air of historical romance. There were endless complaints about the hideous rows of villas springing up around the big cities. A. M. Low was one of many who mocked the British public's desire for mock-Tudor dwellings. He thought the future might lie in structures that could be easily moved from one site to another.[2]

Birnstingl conceded that the interior of some new designs was less cluttered and easier to maintain – like many, he advocated built-in suction cleaners (as long as housewives could be persuaded to abandon their fear of electric motors).[3] The British certainly cherished the notion of an 'ideal home', but it was in America that designers tackled the opportunities most imaginatively. In 1922, *Popular Mechanics* celebrated Walter Dorwin Teague's designs for prefabricated housing. Plastics were touted as the building material of the future. There were plans for rooms defined by roller shutters that could be raised or lowered at will, shallow lakes on rooftops for cooling the interior, and lighting by glass that both transmitted the sun's light and stored it to emit during darkness. Another prediction anticipated power delivered to each house by radio, eliminating the need for wires and hence for static housing. In 1942, it was claimed that all houses would in future be built in sections, allowing the owner to add or remove individual rooms at will.[4]

The anticipation of movable housing gained considerable momentum. In 1935, *Popular Mechanics* reported a statistician's prediction that soon half the population would live in mobile homes that could be towed around by the family car. The most imaginative designs, however, concentrated on entirely new forms for the fixed house. Theodore Morrison's *House Beautiful* of 1929 argued that there would be no true economy of housing until dwellings were mass produced.[5] His words inspired Buckminster Fuller, whose 4-D house of 1927 was later converted to his famous Dymaxion design, a name also used for his new car (Figure 4.1). In 1942, *Popular Mechanics* joined the chorus of enthusiasm when Fuller designed a house that would be built from the top down around a central mast in only a few hours. He called it a 'dwelling machine' to focus on his vision of the home rationally planned around the inhabitants' lifestyle, a view shared by Le Corbusier.

In the end, these predictions for a population on the move were not fulfilled. Totally mobile housing proved impractical and uneconomic as far as the

Fig. 4.1. Buckminster Fuller's Dymaxion house, 1933. Getty Images

construction industry was concerned. In the late 1940s, Britain housed families made homeless in the blitz in concrete dwellings popularly known as 'prefabs', but these were not intended to be moved once erected and were widely recognized as a stopgap until properly built houses became available. There were in any case rival visions for the future of the family home. Ritchie Calder saw the house of the future as being more or less cut off from the outside world, climate-controlled and with no windows. This trend was taken to extremes by some American planners, who envisioned homes moved completely under-ground, a model adopted also by Isaac Asimov in his prediction for 2014. Here they would be safe from both tornadoes and nuclear attacks.[6]

Most architects and town planners shared Low's distaste for new houses built to mimic the styles of previous centuries. They saw the introduction of new materials as a source of inspiration for new designs which would look to the future rather than the past. These would be cleaner in style, lacking all unne-cessary decoration, and would symbolize the move into a better-planned and more efficient way of life. Radical new designs for individual homes were produced and widely publicized for their radical break with the past.

The houses designed by Frank Lloyd Wright in America typified the new styles, with their flat roofs, wide windows and lack of clutter. But these homes were produced for the wealthy who could afford to commission an architect with vision to make a statement about their position in society. Mass-produced houses for the ordinary citizen were much more slowly adapted to new materials and innovative designs. Advertisers increasingly focused their predictions of the 'home of tomorrow' on the gadgets within rather than the house itself. Futuristic styles of architecture were widely presented as symbols of what was to come, but they could be seen more often in commercial and public buildings.

The Rise and Fall of Megalopolis

The most radical alternative to private housing was proposed by the visionaries who saw the whole future of society being transformed by massive building projects organized by the state or commercial trusts. The creation of the steel-framed skyscraper, along with associated technologies such as the elevator, was already transforming the centre of many American cities by the start of the century. The architect G. H. Edgell argued that designs should not hide the internal steel frame and urged his profession to explore new opportunities. In 1928, city-planner Harvey Wiley Corbett predicted skyscrapers half a mile high. Air conditioning was seen as essential for such huge structures and in 1944 *Popular Mechanics* predicted that soon all garbage disposal would be by pneumatic tubes.[7] In his *The Metropolis of Tomorrow* of 1929, Hugh Ferris proclaimed that the drive toward centralized cities was inescapable whatever the concerns of those favouring a more dispersed civilization, although he at least preferred well-spaced skyscrapers. The skyscraper became the symbol of progress, a wonder to behold from a distance – although the cityscape often concealed squalor below.[8]

Europeans looked on in amazement and with some concern. Many feared the materialism of American culture, but others saw both dangers and opportunities. Colonel H. F. C. Fuller visited America from Britain and reported his impression of the vast hustle of New York. He saw America as a child now satisfied with material things, but predicted it would want something more serious when it grew up. Gerald Stanley Lee, another British writer who visited the new world, commented on the crowded humanity of the big cities and saw there the emergence of a new and potentially creative state of mind that would sweep the race forward.[9] Some Europeans worried that their much older (and in some cases much loved) city centres might have to be destroyed to make room for the new order.

Radical architects and town planners such as Le Corbusier did indeed want to tear down the old centres, as had already been done in parts of some continental cities.[10] Others preferred new cities built to create opportunities for a less frenetic way of life. The most radical visionaries went far beyond the demand for more rationally planned cities on the traditional model. They saw the proliferation of skyscrapers as heralding a new world in which the whole human race would become concentrated in huge interconnected structures. Skyscrapers would be linked with walkways at various levels and traffic banished to lower levels to keep vehicles separated from pedestrians.[11] The individual buildings would become so unified that they would eventually be roofed over to become a city totally isolated from the surrounding countryside.

This vision was articulated in some of H. G. Wells's early works. 'A Story of the Days to Come' depicted a world in which everyone lived in apartments within huge cities enclosed in roofs of a glass-like material. Britain has only four such super-cities, with the rest of the land used for mechanized agriculture. There are moving ways with seats for getting about and all eating is in communal cafeterias where food is delivered automatically to the tables.[12] The same image of a huge megalopolis reappeared in *When the Sleeper Wakes*, which Wells presented as an extrapolation 200 years into the future of trends already at work. Again, we have a vision of a single roofed city, all air-conditioned and without windows allowing any view of the deserted world beyond. Energy comes from giant wind-vanes on the city roof, which is pierced to allow access for aircraft to land on the huge buildings below.[13]

The darker side of Wells's vision lay in the society he envisioned within the megalopolis. The poor are confined to the lower regions, in effect transferring the slums of the old industrial cities to the basements of the new. The same image of a society rigidly polarized by wealth was central to Fritz Lang's movie *Metropolis*, made after Lang had visited New York and been impressed by its concentration of skyscrapers.[14] In 1925, the Catholic magazine *The Commonweal* complained of the evil effects of 'The Terrible Super-City', which would produce a dehumanized society with no community spirit and no soul. The 1930 movie *Just Imagine* featured a spectacular backdrop of New York City fifty years ahead, with huge skyscrapers and multi-level transport – although the inhabitants were now known by numbers rather than names and children were produced artificially. The film was not a success, but stills of the city model (which had been built in an airship hanger) were widely reproduced and were used for other science fiction movies. The grandiose designs of Futurist architects such as Antonio Sant'Elia projected monumental buildings melded into a single mechanized entity. They were presented in images devoid of the individual human presence and the movement was subsequently endorsed by the Fascists as an expression of their desire to remodel the human race.[15] In his story 'The Time Machine', Wells had looked even further ahead, imagining a human race polarized by evolution into two species, the degenerate and nearly-blind Morlocks being descendants of the original inhabitants of the levels beneath the great cities.

The idea that city centres might be roofed over did not disappear – it was mentioned by A. M. Low in 1934, although a few years earlier he had predicted underground cities.[16] But Low certainly did not imagine the whole human race confined to a few huge centres, and by this time the whole notion of the megalopolis had begun to seem implausible. Wells himself realized soon after writing *The Sleeper Awakes* that developments in mass transportation were making it easy for people to move away from the city centres. The second chapter of his *Anticipations*, originally published in 1900, was entitled 'The Probable Diffusion of Great Cities'. In 1938, Lewis Mumford defined six stages in 'The rise and fall of Megalopolis', the last two being Tyrannopolis and Nekropolis.[17] The centralized city would eventually impose itself on every aspect of its inhabitants' lives, before self-destructing as its restrictions became unbearable. Eventually, the population would disperse more widely – exactly the trend that Wells and many others had recognized.

The prospect of megalopolis returned in science fiction, but only in response to a new kind of threat. Isaac Asimov's *The Caves of Steel* of 1954 introduces New York detective Elijah Bailey and his robot colleague R. Daneel Olivaw. Set several centuries in the future, Bailey's New York is a huge monolithic structure housing 20 million inhabitants. Everyone lives in apartments which lack not only cooking facilities, but even bathrooms – all food is consumed in public cafeterias and ablutions performed in 'personals' where by convention no one speaks (in the men's rooms at least). Food is produced artificially and transportation within the city is by 'expressways' with strips moving at different speeds. The city is completely walled in (hence the book's title) and is protected by a force shield. It is impossible to see outside and by now most of the inhabitants are psychologically unable to face the prospect of moving beyond the walls. The same fear of the outside grips the inhabitants of Trantor, the capital of the Galactic Empire in the *Foundation* series – although now the megalopolis occupies the whole planet. Arthur C. Clarke also imagined a human race unable to go outside in what is perhaps the most isolated megalopolis in all science fiction, the Diaspar of *The City and the Stars*.

Asimov revived the concept of megalopolis as a response to what he foresaw as a growing shortage of resources caused by overpopulation. With a total population of over 8 billion, the Earth's resources are stretched to the limit and all land is needed to produce raw materials. The crisis lies some time ahead, however, and in his 1964 predictions for the near future Asimov did not mention megalopolis, instead opting for individual homes built underground. Others were not so sure, however. In the same year, Dr L. V. Berkner of Dallas wrote a spoof report from 1984 for Nigel Calder's edited volume predicting the world twenty years ahead (a date already notorious because of Orwell's nightmare vision). He imagined the whole human race concentrated into vast cities of 10 million inhabitants or more. A few years later, Herman Kahn and Anthony Weiner predicted that by the year 2000 the population of the United States would live in only three giant mega-cities. Whatever the details, it was now becoming clear that the drive toward centralization foreseen by many from the start of the century was becoming unstoppable.[18]

From Suburbia to the Garden City

Cities were certainly not going to wither away, but nor were they to be concentrated in monolithic structures surrounded by empty countryside. On the contrary, the countryside anywhere near a major city was increasingly coming under threat from hordes of commuters spilling out into suburbs that sprawled ever further from the centre. Environmentalists and town planners sought to control this dispersal and create a more efficient and attractive way of life. Predictions were made in the hope of shaping the future, presenting what were claimed to be sensible choices between rival technologies. Just as Wells wanted a rational state that would exploit technology for the benefit of all, these visionaries sought to shape policies that would give us the benefits without the negative by-products.

That there were problems in the great cities was only too obvious. Noise was one, and noise reduction was a project dear to the hearts of some enthusiasts. A. M. Low was an expert in this area, while Edward F. Brown of the New York noise abatement movement called in 1931 for the production of noiseless automobiles within ten years. Concerns about pollution were mounting, with efforts to fit 'gas masks' to engines that would remove harmful emissions. Those who could afford it wanted to make use of the new technologies of transport and communication to live outside the city centre. In the 1930s, it was predicted that the suburbs of the great cities would soon spread over fifty or even a hundred miles, and would spread further if personal aircraft ever became available (reviving the scenario of Kipling's Aerial Board of Control stories). As the process developed, the wealthy moved ever further outwards and the need for rational planning became more obvious. Planners such as Lewis Mumford and Le Corbusier had significant influence on local and in some cases national policies for city development.[19]

The relationship between the city as a place for work and the suburbs as a place to live was crucial. Walter Dorwin Teague held that the city centre was not where people should be encouraged to live now that rapid transport allowed easy commuting.[20] The centre would necessarily be a location for concentrated activity, and this was where high-rise buildings were essential. Yet simply replacing old structures with new ones in the existing locations would fail to address the problems of congestion and pollution that had arisen within the old, unplanned city centres. Radical planners such as Le Corbusier argued that the old centres had to be torn down and replaced with rationally ordered layouts that would permit maximum efficiency. In his *The City of To-Morrow* of 1929[21] he celebrated the power of science and technology to transform our lives and insisted that: 'A town is a tool.' His 1922 plan for a city of 3 million people envisioned skyscrapers 220 metres high with all traffic carried below ground and a huge central station (with an airport on the roof) to allow rapid commuting. The larger buildings would be cruciform or Y-shaped to allow maximum access to light, or be stepped inwards like the ancient ziggurats of Babylon. Christian Barman's contribution to the 'Today and Tomorrow' series pointed out that having aircraft runways on the roof didn't fit with designs for pyramidal skyscrapers.[22] The hope of having airports on the roof soon faded (until the age of the helicopter), but the idea that the larger building should be more spread out remained popular. If the planners had their way, the crowded skyline now characteristic of the American city would become a thing of the past. As a magazine editor said when confronting Le Corbusier's 1922 plan (dubbed 'The City of the Future' by journalists): 'In two hundred years Americans will be coming over to Europe to admire the logical productions of modern France, while the French will be standing in astonishment before the romantic sky-scrapers of New York.'[23]

Not everyone welcomed the destruction of historic city centres. One solution suggested by Percy Lockhart-Mummery was to preserve the old as a monument and build the new city alongside. He described a fictional visit by a family from New Zealand to the London II of 2456, a modern megalopolis built to the south of the historic London I (Figure 4.2).[24] Most planners looked for alternatives to the sprawling suburbs that were spreading out from the city centres. Even Le Corbusier had accepted that people would prefer to live in low-rise blocks surrounded by greenery in the suburbs. They might prefer individual houses, each with its own garden, but as Ritchie Calder insisted, this was not a realistic option – the new planned suburbs would have to consist of 'island' buildings set amid parkland.[25] But *was* it unrealistic? Planners such as Lewis Mumford were increasingly concerned about the dehumanization of life in the ever-larger cities and saw the preservation of the family within its own home as vital to psychological health. Mumford's insistence that the megalopolis was doomed to self-destruction was coupled with plans for rationally ordered smaller towns. His *The Culture of Cities* of 1938 and the later *City Development* show his commitment to the movement that offered the most extreme alternative to megalopolis: the campaign for garden cities.

Mumford acknowledged the influence of two British pioneers in this area, Patrick Geddes and Ebenezer Howard. Geddes and Mumford only met twice, and neither meeting was a success thanks to the former's over-enthusiasm and the latter's desire to forge a literary career.[26] But the two corresponded extensively, and Mumford recognized the potential in

LONDON A.D. 2536

Fig. 4.2. London as it might appear in A.D. 2536. Frontispiece to
J. P. Lockhart-Mummery, *After Us* (1936).

Geddes's call for an evolutionary model for city development that would
create an environment in which human nature could achieve its full potential. In books such as his *Cities in Evolution* of 1915, Geddes proclaimed
the need to get rid of the slums that stunted people's development, replacing them with better housing in more dispersed locations. Mumford
provided an introduction to a 1946 reprint of Ebenezer Howard's *Garden
Cities of To-Morrow* of 1902. Here, Howard proposed an alternative to
suburban expansion, calling for the creation of entirely new centres of
population close to the big cities but built on open land that would allow
rational planning of roads. People would live in a natural environment –
there would be a central park surrounded by a radial grand avenue and
well-spaced streets. On this model, the private home could survive and give
everyone access to open space. The new houses would be built with the
latest materials and include all modern conveniences.[27]

    The planners who called for garden cities were going beyond prediction –
they were seeking to direct the course of social development along lines driven
by human needs rather than technological possibilities. Howard's campaign did

indeed motivate the creation of garden cities such as Letchworth and Welwyn near London and Raburn, New Jersey and Greenbelt, Maryland. His views were influential both in the New Deal and in British plans for development after World War II. But in the end, practicalities ensured that full implementation of his vision was restricted. As Calder had insisted, there was never going to be enough room for everyone to have a cottage garden. For most visionaries, it became increasingly obvious that even if new cities were to be planned, they would have to involve construction on a large scale. The planners' aim was to ensure that the ensemble would preserve human proportions while presenting an image of future progress.

The Streamlined City

Lockhart-Mummery's 1936 vision of London II was certainly grandiose, but the huge buildings were imaginative and by no means monolithic, and they were set in surrounding parkland.[28] The architects of the early twentieth century were keen to display their rejection of tradition by creating designs that – while often too visionary to be built – were symbolic of the drive toward a new world. The plans were sketched in images presented to the public in books and magazines and in the many expositions and world fairs that tried to encourage interest in new technologies. There were certainly rival ideologies underlying their work, but all sought to use their vision of the future city as a means of expression.

The insistence that innovative designs were needed in response to both new technologies and new social demands became active from the start of the century. Before the Great War, architects such as Frank Lloyd Wright in America, the Bauhaus school in Germany and the Italian Futurists had begun to design, and in some cases build, domestic and larger structures that marked a radical break with the past. The Futurist Antonio Sant'Elia proposed his Citta Nuova in 1914, imagining an impersonal megalopolis that symbolized the move toward a rational social order, a vision later taken up by the Fascists. Other totalitarian regimes also used grandiose architecture to embody their vision for the new, all-powerful state. Less deliberately overwhelming, the International style which developed from the Bauhaus designers embodied a severely practical approach which was nevertheless frequently associated with utopian rhetoric. This vision was proclaimed in the 1937 Exposition international des Arts et des Techniques appliqués à la Vie moderne. Its severe designs were created for factories and department stores, but were soon being applied to the apartment blocks into which the inhabitants of the new cities would be moved.[29]

American architects and planners also imagined huge impersonal blocks, as visualized in Hugh Ferris's *The Metropolis of Tomorrow* of 1929. But soon they began to move in a different direction. Paul T. Frankl celebrated long, horizontal lines as an embodiment of speed,[30] but the prevailing style that emerged used curves to symbolize the smoothness of passage through the air. Even when applied to static buildings, the style known as 'Streamlined moderne' was meant to imply the rapid transition to a better future. As developed by designers such as Norman Bel Geddes, this style shaped the appearance of both interior and exterior forms. This was no vague or distant utopia: it was a future toward which we were all moving with amazing rapidity, led by the commercial interests that were going to create the hardware for the new world. In the New York World's Fair of 1939, this corporate vision of the future was presented as the route-map that would guide us from the Depression into a better world. It was the world of tomorrow, not a vague hope for utopia, and it remained a brighter vision of the future as the International style's monolithic apartment blocks became the symbol of impersonal bureaucracy.

The 1939 World's Fair exhibited several versions of corporate America's plan for the future. Inside the Perisphere was Henry Dreyfuss's model of Democracity, billed as the perfectly integrated garden city of 2039. It had a huge central skyscraper, surrounded by circular avenues feeding the garden suburbs. The General Motors pavilion had the hugely popular Futurama model of Norman Bel Geddes's ideal city for 1960 – well spread out, but completely dependent on the car as a means of transport (see Fig. 1.1, p. 13). It provided an effective image of a pleasant future environment for all, but the underlying message was pinpointed by Walter Lippman, who wrote: 'GM has spent a small fortune to convince the American public that if it wants to enjoy the full benefits of private enterprise in motor manufacturing, it will have to rebuild its cities and highways by public enterprise.' The Chrysler pavilion had a model of Raymond Loewy's design for a future city based on a Rocketport for transatlantic travel. H. G. Wells visited the fair and was told that it showed the public the brighter face of what the future would bring.[31]

The fairs were only the most concentrated source of futurist imagery. Although most of the wilder plans sketched by architects were never built, their new styles did gradually begin to percolate into the real world. They were frequently used by governments and corporations to advertise their commitment to progress. The International style was applied to factories from its very inception at the Bauhaus and its severe, block-like formations soon began to appear in remodelled cities and suburbs. In the 1950s, new industries such as atomic power plants were designed to have what was now recognized as a futuristic appearance, although the effect was often more intimidating than appealing. Geddes and other proponents of Streamlined moderne also designed factories and public buildings ranging from theatres to gas (petrol) stations, and again this rival vision seems to have chimed more effectively with what the public hoped the future would look like. In America, grandiose plans for totally new cities were still being proposed in the 1950s, as in the hypothetical Tottenville described in *Popular Mechanics*. This was to be a suburb with a population of a 100,000, with lots of parkland, no pollution and two-decked roads.[32] The downside of the transition to road transport promoted at the 1939 Fair only became apparent in the 1960s when the expressways originally built to get city-dwellers to the countryside and seaside began to destroy existing communities. The disastrous effect of Robert Moses's Cross-Bronx expressway in New York was a prime example.[33]

Europe only gradually recovered the confidence to look forward at the end of World War II. When it did, futuristic architecture again became the symbol for a vision of a better world based on applied science and technology. A good example of this symbolism can be seen in the 1951 Festival of Britain, where the Dome of Discovery publicized the latest developments. The Festival certainly did look forward, but some historians have noted that it also reflected a deep sense of nostalgia for traditional values.[34] This ambivalence can even be seen in the juvenile literature of the time – in the space-flight feature 'Dan Dare' in the comic *Eagle*, futuristic cities reminiscent of the designs seen at the World's Fair were depicted alongside a traditional rural world where cricket was still played on the village green. *Eagle* was published to counter the materialistic values emanating from America, and its attitude to science and technology combined a thirst for progress with glances back to the past.[35]

The tensions that had emerged in the early part of the century were never resolved. Planners wanted suburban apartment blocks, but the new social environment they created was more popular in some countries than others. In Britain, the tower blocks into which families were crammed against their will became a symbol of misguided planning and many have now been torn down.[36] Cities have certainly continued to grow – but they have grown both upwards in the centres and outwards on the peripheries. Fantastically designed skyscrapers now tower over the older cities and completely dominate the newer centres of the developing world. They are often linked together by air-conditioned enclosures, especially in regions with severe climates. The tendency to provide a unified artificial environment is also visible in the omnipresent shopping malls of the suburbs which sprawl ever further outwards. In the end, many of the predictions made in the first half of the century were realized in part during the second, social forces having made room for several competing new technologies to display at least some of their potential.

5 Communicating and Computing

Better means of communication would have a major influence on future developments. If instantaneous communication with individuals or whole groups became possible, the need for humanity to huddle together in cities would diminish, paralleling the effects of rapid transportation. Major changes in communications technology had already taken place by 1900. The telegraph system now transmitted messages instantaneously around the world. Radio was soon providing an alternative that allowed messages to be sent even from ship to ship. A telephone system already existed in most big cities and it was assumed that it too would be extended to long distances. The cinema allowed film makers to transmit ideas and entertainment to audiences around the world. These existing technologies provided a template for speculation about potential developments and a warning that further progress would have social impacts far beyond those already apparent.

Most predictions assumed further improvements in what was already available, or obvious combinations of existing systems. Telephones would soon enable us to talk to anyone in the world, and combined with radio would become portable. Telegraphy and radio would be able to transmit speech and even pictures. Cinema would have sound as well as coloured images. Putting the whole package together would give television – the instantaneous transmission of moving images. All of these predictions abounded in the popular press and in the growing world of science fiction. Yet there were surprising blind spots, some revealed when ordinary people played a role in showing what they wanted from new technologies. Marconi saw radio only as an extension of telegraphy – the potential for deliberately broadcasting to an extended audience was only recognized in the 1920s. The majority of movie makers initially thought there would be no public demand for the 'talkies'.

The most obvious blind spot in retrospect was the failure to anticipate what have become the most important products of the new communications technology: the computer and the internet. Punch-card computing did have some impact on business, but few realized that the technology being developed by the communications industry would extend the power of the computer to every field of life. Even when the first electronic computers were built in the 1940s, no one foresaw the emergence of the home computer. It would be easy to say that this was merely because the transistor was not yet available – but the robots already imagined by science fiction authors (including Asimov) would have required computers small enough to fit inside a skull-sized space. Nor did anyone imagine that by linking computers together, information could not only be disseminated from a central authority, but could be uploaded by anyone with access to the system. Here, the model provided by the broadcasting industries presented a barrier limiting the imagination even of those who were creating the new technologies.

Whatever the limitations of their visions, where the prophets thought they could see the way ahead, they were well aware of the effects the innovations would have on the way in which people live. The inventors and those who promoted their work were obviously enthusiastic – although the interactions between them were often fraught with difficulties. It was easy to imagine the short-term benefits of television or the mobile phone. Who would not want to receive news and entertainment visually in the home, or talk to a friend while on the move? It was also easy to pretend that the wider implications of the new technologies would necessarily be benign. Cinema, radio and television would boost the interactions between communities and promote world peace, as well as being of obvious benefit to education.

The pessimists, too, had a field day imagining the downside of the revolution to come. The various media could fall under the control of ambitious politicians or ruthless capitalists who would use them to condition the population to their will. People would become so used to the passive consumption of entertainment that they would no longer have a real life. Few noticed that the ways in which technology was used fluctuated through time. In the era of the radio hams, the new medium was empowering for individuals, but the power was soon swept away as government and industry took over. The story of how new technologies emerged in this field provides a classic illustration of how big corporations were increasingly able to exploit both the individual inventors and the public. Paradoxically, the internet seems to have had the opposite effect and the computer, introduced as a means of control, has allowed everyone to have their say once again.

Moving Pictures

All of the new technologies transforming people's lives in the first decade of the new century were ripe for further development. The telegraph, by now well established, offered the possibility that it might be able to transmit pictures as well as Morse code. The telephone was already widely available, but was confined originally to large cities and dependent on human operators. The phonograph or gramophone provided recorded sound, but the quality was poor and reproduction was at a low volume due to lack of amplification. The silent cinema was already a significant feature of mass entertainment, but there was the obvious possibility of adding a soundtrack and of photographing in colour. Inventors were at work on a whole range of improvements, some eagerly anticipated by industry and the public, others more visionary. The most exciting prospect was an extension of the other great miracle of the age – radio – to include the transmission of images. Television and telephones with visual as well as audio connections were eagerly anticipated, but would prove hard to develop in practice.

The science fiction authors were the most imaginative and looked furthest ahead. In *The Sleeper Awakes*, H. G. Wells pictured his hero puzzling how to operate a system in which both sound and vision are recorded on cylinders to be played back at leisure in the home. Hugo Gernsback's 'Ralph 124C 41+' of 1911 has a tele-theatre that reproduces live performances and a 'telephot' which is a video-telephone (with a translator for good measure). The latter innovation also featured in his magazine *Radio for All* in 1922. A decade later, Wells suggested that the problems delaying the creation of workable television would soon be overcome, but saw the main application being for personal communication rather than broadcasting. As a result, '[t]he whole world will become a meeting place'. Over the next few decades, science fiction authors took it for granted that these advances would be forthcoming, although in the real world advances were slow. *Popular Mechanics* described the picture-phone as still experimental in 1957, although in 1950 it had predicted that in fifty years' time the housewife of Tottenville would do her shopping by this means.[1] The videophone connection from a space station to Earth in the movie *2001: A Space Odyssey* was still an effective piece of futurology in 1968.

Popular science writers tended to be more sober in their predictions. There were numerous accounts of the telephone system being extended on an international scale and of automatic exchanges that would replace the operator and thus speed up connections. Everyone assumed that methods for recording sound would improve, but once the disk record had replaced the cylinder, higher levels of fidelity became the main goal. One potential technology that did catch the public's attention was the extension of the humble telegraph to allow the sending of pictures – a relatively simple application involving the transmission of pixel-style information to a printer. In 1929, A. M. Low predicted what would become the fax system of the late twentieth century, while a decade later, Lee De Forest imagined newspapers printed by such a system in the home. Fax would be a boon to industry and extending the process by radio transmission would make it international in scope.[2]

The telephone and the phonograph became more widely available in the United States, although they remained something of a luxury in Europe. Cinema had the most widespread impact on popular culture. By the 1920s, the movie industry was well established and most city-dwellers would go to a cinema regularly. To the inventors such as Edison, it was obvious that there were two major ways to improve the silent movie: the addition of sound and colour photography. Whatever the system created to record and synchronize sound, the 'talkie' remained impractical until loudspeakers became available in the 1920s. By then, several groups were working on the problem and the first short talkies were produced by Warner Brothers in 1926. Despite much skepticism in the industry, they were quite successful, and in 1927 *The Jazz Singer* famously pioneered the full-length talkie and became an overnight sensation. The skeptics were forced to recant and soon the era of the silents was over.

The reluctance of many movie producers to imagine that the addition of sound would be so attractive illustrates the complexity of the prediction business. Their doubts reflected a cultural inertia – the public had got used to the silents and were fascinated by their stars, so why rock the boat? Whatever the interest among the inventors, it is significant that the British popular science magazine *Conquest*, published from 1919 to 1926, did not print a single article anticipating the coming of sound. Its successor, *Armchair Science*, provided copious details of how the talkies were produced, reflecting the public's interest in the new technology once it had become available. The media skeptics' viewpoint may thus reflect a genuine blind spot in the public's awareness: they liked the silents and just didn't appreciate how much sound had to offer until they actually experienced it.

If *Conquest* did not anticipate the talkies, it did print an article hailing an attempt to develop a system to introduce colour in 1924. It would be the end of the following decade before the process was actually perfected (*Robin Hood* with Errol Flynn was issued in colour in 1938). *Conquest* also reported a new process for producing film on plastic, noting that it offered the benefit of allowing artificial images to be used as background. Its comment represents a striking anticipation of effects that would only become realistic in the age of computers: 'The film plays of the future may utterly sweep away the plays in which real people are engaged to act upon the stage, for the film can bring to the stage scenes which could not be possible otherwise.'[3]

The addition of sound and later colour was making the movies even more attractive, but as the last quotation suggests, there were wider consequences to consider. Would the movies drive the ordinary theatres to extinction? Television would later be accused of having the same effect on the cinemas. Social commentators raised more serious issues, sensing that the availability of powerful alternatives to real life would create a generation of passive consumers no longer able to deal with the real world. For Georges Duhamel, this was all America's fault: Hollywood was having a stupefying effect because people exposed to too many movies became incapable of doing anything requiring resolution or thought.[4] The more realistic the entertainment, the greater its ability to replace real life, a threat already anticipated in the world of E. M. Forster's 'The Machine Stops'. In *Brave New World*, Aldous Huxley added another layer of sensation with the 'feelie'. Here, the all-round artificial world of entertainment was used to pacify and enhance conformity. Olaf Stapledon went one step further in his *Star Maker* of 1937 by imagining a humanoid race reduced to impotence as the whole population is ensnared by 'radio bliss' – sensations transmitted directly to the brain.[5] Three years later, the science writer Waldemar Kaempffert predicted the transmission of touch and taste.[6]

Radio Wonders

Many of the developments envisioned for communicating images depended for their implementation on the advances made in the radio industry. Radio had become a popular sensation at the beginning of the century as pioneers such as the British physicist Oliver Lodge and the inventor Gugliemo Marconi struggled (and competed) to develop a workable system for transmitting messages 'over the ether'. Marconi became famous by operating the most successful system, but he always envisioned radio as merely a new form of telegraphy without wires – hence the British term 'wireless' and the French 'télégraphie sans fil', which remained popular into the middle decades of the century. When it was installed on ships, the primary purpose was to allow passengers to communicate ashore and for Marconi a crucial concern was to preserve the confidentiality of the messages. In times of crisis, the ship's radio could send a distress message that could be received by any vessel or shore station with the appropriate equipment. When the White Star liner *Republic* out of New York sank in January 1909, its radio operator Jack Binns was hailed as a hero and Marconi basked in reflected glory (although radio's role in the *Titanic* disaster was more controversial).[7] The early equipment had major limitations, but everyone could see the potential for further development.

It was the amateur enthusiasts or 'radio hams' who realized the true scope of the new technology.[8] For a generation of young men around 1920, building and operating your own radio set was the new symbol of masculinity – a new frontier to conquer. Eventually, governments (aided and abetted in the United States by business interests) would restrict the amateurs' activities in the name of safety and security. But in the meantime, there was a wave of enthusiasm for all applications of radio. For the amateurs, the whole point of operating a set was that you could hear what everyone else was saying, and they could hear you. They became the true pioneers of broadcasting, soon to be overtaken by centralized operators such as the Radio Corporation of America (RCA) and the British Broadcasting Company (later Corporation, the BBC). Sending out news and entertainment was not the only application of the new technology. For enthusiasts such as Hugo Gernsback, radio had endless potential uses, all hailed in magazines such as *Radio for All* and his science fiction pulps. Britain had the Wireless Press, which published the magazine *Conquest* and an array of specialist magazines and books aimed at the radio enthusiasts. A. M. Low's *Wireless Possibilities* of 1924 was the 'Today and Tomorrow' series' contribution to the wave of enthusiasm.

For the British, the setting up of a long-distance radio network for their overseas Empire would be a significant improvement over the old telegraph system. To make such a system effective, international agreements to limit interference between stations would be necessary. Europe led the way here with a Broadcasting Bureau established in Geneva – and America soon had to follow suit.[9] Many argued that navigation for ships and aircraft would be improved by a worldwide system of radio communication (although no one foresaw the technique that would become radar).[10] There was endless fascination with the possibility of radio control for aircraft, ships and even land vehicles (including tanks). These accounts often had to warn readers that the radio only transmitted instructions to the vehicle – the popular expectation that power itself could be transmitted through the ether was not a viable proposition, even though it had been suggested earlier by figures as prominent as Nikola Tesla. At the turn of the century, he built a tower at Wardenclyffe, NY, intended to broadcast power from Niagara Falls, although few readers of the popular literature announcing his plans realized that in fact they depended on electromagnetic waves transmitted through the earth (Figure 5.1).[11]

One of the hottest topics was the prospect of connecting telephones by radio and reducing their size so that the mechanism became portable – what we now call the mobile or cell-phone. There were early expressions of hope even before the Great War, but the theme only became popular during the radio boom of the 1920s.[12] In his *Wireless Possibilities*, Low predicted that in a few years' time it would be possible to talk to a recipient anywhere in the world, even when flying on an aeroplane. Five years later, he made a similar point in one of his regular *Armchair Science* features: 'I shall be glad when we have made wireless sufficiently selective to enable me to ring up during every rail journey I make and talk direct to my friends.' Note that his concern was the problem of interference between transmitters, not miniaturization. He also recognized that there would be a downside to the facility: 'Why should I inflict a description of my mother's children to a radius of six yards, until all those around are driven to fury ... ?' Low thus not only predicted the mobile phone – he realized what a nuisance they could become when used in public. Other commentators remained enthusiastic at the prospect through the 1930s, including I. O. Evans and C. C. Furnas.[13] Furnas thought a vest-pocket-sized phone would be particularly useful for hunters, surveyors and the police.

Fig. 5.1. Artist's depiction of Nikola Tesla's radio tower at Wardenclyffe, NY, from which messages and power would be broadcast. Alamy images

These authors assumed that it would only be a matter of time before radios were developed that would be small enough to fit into a pocket. The thermionic valves that became widely available in the 1920s made the transmission and amplification of sound feasible, opening up the prospect of radio telephones and the broadcasting of news and entertainment. But they were bulky and produced considerable heat, so they were not suited to miniaturization on the scale required for a mobile phone. Portable radios were produced as early as the 1930s, but they were only 'portable' in the sense that a person could just about carry one. Low and his fellow enthusiasts simply assumed that some new technique would be developed, but they had to wait a long time. In his *Electronics Everywhere* of 1951, Low was still predicting the mobile phone, but now he placed its development in the far future.[14] Curiously, his book makes no mention of the invention that would eventually make it possible – the transistor. This was based on effects discovered in the late 1940s, but to be fair to Low its applications only became apparent in the years immediately following the publication of his book. Its developers, John Bardeen, Walter Brattain and William Shockley, shared a Nobel Prize in 1956. It would be the 1960s before the life of the ordinary person would be affected, and then it was the transistor radio that had the most immediate impact.

Broadcasting

Efforts to transmit sound by radio began at the turn of the century. It was only in the 1920s that it became feasible to broadcast voice and music to a large audience. The radio hams of the early 1920s had exploited the potential of the medium to communicate widely with one another, but the result was a chaos of interference that forced governments to impose restrictions. By the mid-1920s, it was becoming clear that centralized broadcasting was the way forward, with the stations supported either by commercial or government interests.

Conquest hailed the potential with an editorial in 1921. It suggested that it was hard to see where the movement would end, and the effect on other entertainment media would be unpredictable. Politicians might use it to reach the public, but 'one shudders to think of the awful possibilities of etheric advertising'. By 1924, *Popular Mechanics* was telling Americans that it would soon be possible to broadcast to the whole continent via a series of powerful relay stations. It too noted the potential for politicians to address the nation.[15] It was the Americans who got the advertising that *Conquest* had feared, while the British opted for a government-supported system, which became the BBC. This corporation was set up to ensure that the service would provide news and educational material along with entertainment. For some time, the public complained about the lack of popular music and other light entertainment. Birkenhead predicted that by 2030 a huge amount of entertainment would be available in the home, although he thought that people would still go to the theatre and opera for the sense of occasion.[16] Other European countries also opted for stations under government control, although the downside of this system soon became apparent with the rise of Fascism.

The anticipations of the early 1920s recognized that broadcasting might be a mixed blessing and its effects unpredictable. As early as 1912, Marconi argued that radio would precipitate a move toward a more centralized world and make war impossible. In 1928, Philip Gibbs thought broadcasting would transform our habits by altering our perceptions of time and space.[17] There were plenty of optimistic commentators. Writing in *Armchair Science*, Fournier d'Albe thought the wide distribution of information by radio would promote internationalism, limited only by the problem posed by different languages. Anticipating the move to television, the magazine again saw a boost to internationalism as people gained a better perception of the wider world. I. O. Evans took this one step further by predicting that radio would boost the move toward an international language.[18]

St Barbe Baker saw broadcasting as a means of extending education to remote regions, especially in Africa. The same theme was taken up by Lee De Forest, the American inventor of the thermionic valve:

Broadcasting must aim to mould future civilization along lines of greater knowledge, finer culture, and a broader understanding. We are tearing down former barriers. For the humble workman's cottage and the millionaire's mansion enjoy alike the benefits of radio. It is the democritization of music, of the histrionic art, of the banquet-hall, of the political rostrum, and of culture at large. We aim to apply radio in our schools, so that our leaders, in every field of achievement, may appear in many schools to thousands of pupils at one time.

Some pushed the sense of optimism even further. Addressing the American Association for the Advancement of Science as late as 1938, the physicist Arthur H. Compton still saw the progress of communications as something that would integrate the global community so that 'the world becomes also a conscious unit, very similar to a living organism'.[19]

Negative consequences were also predicted. In Britain, there was concern that the provision of entertainment in the home would accelerate the decline of religion, even though the BBC broadcast church services.[20] Radio entertainment also played into the fear of an ever-more passive population. Just as they sat in cinemas to absorb the movies, now they would sit at home to listen to popular music. Low, always vigilant on the subject of noise pollution, warned that if broadcast music ever became available through portable equipment, rail travel would be unbearable unless everyone used headphones.[21] As with the mobile phone, he realized that people would use the new technology without thought for the comfort of their neighbours.

Pessimists worried about the power that might be wielded through the new technology by advertisers and politicians. George Orwell was by no means the only commentator to imagine how an autocratic government could use the radio (and later television) to maintain its control through the dissemination of propaganda. Exactly this point was made by J. N. Leonard in 1935, citing the dictatorships of Europe (with the BBC as an honourable exception). He also objected to the commercialized American system as 'a sort of multi-tongued huckster'.[22] The panic following Orson Welles's classic broadcast of *The War of the Worlds* on 30 October 1938 showed how a well-managed radio programme could manipulate the popular imagination.

The fact that so many people could be fooled into thinking that the Martians were invading reminds us that many still believed that the planets might harbour some form of life. From the earliest days of radio, the possibility was mooted of either receiving signals from Mars, or sending them ourselves. At one point, Tesla wondered if the interference picked up by radio receivers might be emanating from Mars. Low wrote a science fiction novel *Mars Breaks Through*, in which a Martian takes over the mind of a human financier, ostensibly to save our world from war, but actually to build weapons that will destroy his own enemies.

There was an important twist in Low's plot, however – Murchison, the financier, had left himself open to mental control because he had built a machine to amplify his telepathic ability enough to reach Mars. In his popular science works, Low also promoted the view that telepathy was a real phenomenon allied to radio, with the human brain acting as sender and receiver. The same idea was promoted to younger readers by I. O Evans.[23] Radio had been connected with the paranormal from the beginning, often portrayed as 'magic' in the popular press. The link seemed all the more plausible since Oliver Lodge was both a pioneer of radio and a leading promoter of spiritualism. For him, the spirits of the departed continued to function in an ethereal world, imprinted on the medium that was still popularly supposed to transmit radio and other electromagnetic radiation. In February 1937, Lodge persuaded the BBC to conduct an experiment to test the validity of telepathy. The radio alerted people around the country when 'senders' were trying to project images for them to detect telepathically.[24] The writings of Low and Lodge created an expectation that what would later be called the paranormal was just an extension of the real world studied by physics. Radio provided a model which seemed to offer the hope of understanding and controlling real functions within the human brain.

Television

Even before the broadcasting of sound radio was achieved, there had been predictions that vision, too, would eventually be added. The name adopted almost immediately for this hypothetical technology was 'television'. Popular science writers followed up these early hints, monitoring and anticipating the efforts of inventors and industrialists with varying degrees of optimism, offering a classic example of prophecy driving or at least encouraging innovation. There were also blind alleys that seemed plausible at the time, including the expectation that the future of television lay in transmission to cinemas. Through the interwar years, pioneers such as John Logie Baird and Filo T. Farnsworth struggled to develop a practical system for transmitting moving images and the first broadcasts began in the years immediately before the outbreak of war. The first sets were tiny and the images blurred, and it was only in the 1950s that television began to transform people's home lives. Not everyone thought it was worth the wait.

Science fiction writers took the transmission of moving images almost for granted, but some of the earliest suggestions in the popular press emerged from extensions of an old technology – the telegraph. We have already seen how popular science writers promoted the idea that the telegraph could be used to transmit images. Several early predictions of television saw it as little more than a speeded-up faxing system. If one could fax a sequence of images quickly enough, the recipient would see changes happening almost in real time, if a little jerkily. Low's *Wireless Possibilities* floated the idea of such a system in a chapter entitled 'Radio Television', and two years later Lee De Forest made the same prediction in *Popular Mechanics*, again using the term 'television'.[25] This was a false start, however, and already efforts were underway to develop a means of transmitting images to a screen so rapidly that they would appear to move.

Early in 1927, the *New York Times* hailed a demonstration of television at the Bell Telephone Laboratories, but the quality was so poor that the project was abandoned.[26] In 1929, the first issue of *Armchair Science* explained the latest developments and hailed future prospects, with several equally optimistic articles following two years later. T. Barton Chapple concluded in terms illustrating the extent of popular interest: 'small wonder that on every side the opinion is expressed that television will bring about a new era of every-day life'. In 1931, John Logie Baird visited New York, where he met H. G. Wells and gave a radio talk offering an equally optimistic prospect:

Throughout the world the highest scientific thought is being devoted to television. Vast strides have been made and will be made in this new art. I myself look forward to seeing at no distant day television theatres supersede the talkies, and the home television become as common as radio is today.[27]

Baird formed a company that year to promote his mechanical system, but it was becoming increasingly obvious that the future lay with the alternative based on the cathode-ray tube.

Baird foresaw television replacing movie films in the cinema as well as smaller-screen versions for the home. The popularity of the movies was so great that many commentators envisioned big-screen television as the way forward, oblivious of the fact that inventors struggled to get images transmitted to even tiny screens in their workshops. Cinema television was the future predicted in Low's *Wireless Possibilities* and in Barton Chapple's articles cited above. *Popular Mechanics* was still promoting it in 1944 and after the war – by which time home television was a reality – Low was still lamenting that the technology did not allow it to operate in cinemas. Ritchie Calder's *Birth of the Future* took the opposite tack, predicting that television cameras would never be good enough to transmit live action, so the future lay with broadcast movies.[28]

In 1936, C. C. Furnas still thought that a workable system of television lay a considerable time ahead, yet by the following year *Armchair Science* was eagerly anticipating the BBC's first broadcasts. The magazine's monthly 'World of Wireless' series frequently provided details of what was expected. In 1936, the French magazine *Science et Monde* proclaimed that advances were being made so rapidly that television would soon become commonplace, helping to undermine our sense of global distances. It reported a demonstration at an exhibition in Brussels and reminded its readers that the technology had been predicted by French writers from Jules Verne onwards.[29] By August 1937, there were announcements of the first BBC trials, German broadcasts from the Olympic Games and RCA's plans for demonstrations at the forthcoming New York World's Fair.[30] The outbreak of war brought these early schemes to a halt, and it was only in the late 1940s that television once more began to move ahead. In the course of the 1950s, black and white televisions slowly began to replace radio as the main source of at-home entertainment.

Predictions now switched to anticipating improvements in the system. Colour television was predicted as early as 1944 in *Popular Mechanics*. Kaempffert thought there would soon be stereoscopic images and the transmission of touch and smell (making Huxley's 'feelies' a reality). Soon every aspect of a person's behaviour could be recorded to provide a kind of immortality. *Popular Mechanics* pointed out, though, that there was an important technical barrier to the long-range transmission of television signals: the short waves used allowed line-of-sight communication only. It suggested a huge network of relay stations to allow national coverage. Arthur C. Clarke's recognition that satellites in geostationary orbits could provide global television coverage was published in *Wireless World* as early as October 1945 and popularized in his *The Exploration of Space* in 1951 (Figure 5.2).[31] It would be some years before the two technologies at last came together.

From the start, there were commentators who recognized that the popular enthusiasm for television deflected attention from its darker possibilities. George Orwell, already well aware of the potential of radio for the spreading of propaganda, foresaw the far more pervasive effects of television in his *1984*. The fact that the 'telescreen' of the novel worked both ways, allowing the state

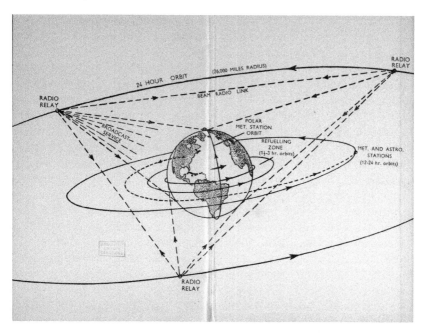

Fig. 5.2. Arthur C. Clarke's proposal for communication satellites in geosynchronous orbits. Rear endpapers of Arthur C. Clarke's *The Exploration of Space* (1951).

to brainwash its citizens and to spy on them, made the prospect even more forbidding. As the use of television spread in the 1950s, debates over its social impact intensified. Some welcomed its value in education, others argued that even more than cinema and radio it would create a generation of passive consumers. In 1964, Sir Gerald Barry of the United Kingdom's Granada Television predicted that by 1984 twenty-four-hour news coverage would eliminate the newspapers, while televised entertainment would make books redundant. He also thought that documentary programmes made on a global scale would reduce people's desire to travel around the world.[32] The modern world has seen some of these predictions fulfilled, while others seem increasingly wide of the mark.

Computers

If television and the mobile phone were anticipated decades before they were actually created, the same cannot be said of what is now seen as the most crucial application of electronics: the computer. Radio technologies had flourished with the invention of the thermionic valve and were miniaturized in the 1950s using a second innovation, the transistor. Electronic computers used the same technologies, but neither their initial creation nor the miniaturization that led to the home computer were anticipated in the pre-war years. Here is perhaps the clearest illustration of the fact that predicting the future of technology can have blind spots. Some future developments seem obvious long before they are feasible with the knowledge available at the time, while others are so far outside the range of imagination that they are simply inconceivable. Even the science fiction writers of the interwar years did not imagine a world where machines would have the capacity to take over a wide range of human functions simply by doing calculations rapidly and by storing the resulting information. Computers arose from inspired responses to military necessity in the war years and were only later recognized as having wider uses (although a few pioneers suspected this potential from the start). Even then, the prospect that they could be reduced in size to allow everyone to have their own machine was ignored, although the miniaturization of communications technology had been anticipated long before the transistor radio and the mobile phone were developed.[33]

Nowhere is this failure of the imagination clearer than in science fiction. Reading the work of the pioneers, one is struck by what now seems an incongruous mix of advanced technologies – some so advanced they still have not been developed today – functioning alongside older systems that we know were soon to be swept away. Rocket ships and ray guns are employed alongside slide rules and printed tables of information.[34] E. E. 'Doc' Smith's classic space operas offer a prime example. In his *The Skylark of Space* series (the first volume of which appeared in 1928), the hero develops space travel on an interstellar scale while still measuring equipment with calipers and doing calculations with a slide rule. Asimov's *Foundation* series, written in the late 1940s, has a galactic empire that still uses newspapers and metal coins and where the projected *Encyclopedia Galactica* will be deposited in libraries as though in book form. Robert A. Heinlein's *Starman Jones* of 1953 has inter-stellar travel for which the ship's course is plotted by human astrogators using information from printed tables. They do have computers, but these are little more than calculators. The earliest story that has been hailed as a partial anticipation of the internet is Murray Leinster's 1946 'A Logic named Joe'.[35]

The absence of powerful computers and most forms of information technol-ogy from these stories is striking. A rare exception is Lawrence Manning's novel *The Man Who Awoke* of 1933, which imagines a future in which the Earth is governed by a giant computer known as 'The Brain'. Popular science works from the interwar years occasionally reported the use of mechanical calculators and punched-card systems for information retrieval, but saw little beyond their obvious value to large institutions. The first electronic computers were developed for military purposes during World War II, pioneering work being done by mathematicians such as Alan Turing, who used them for code breaking. This work remained secret long after the war, thus obscuring the true origins of the technology. Military origins were acknowledged by Low in a chapter on 'Electronic Brains' in his 1951 book – but the applications noted were for gun-laying and aircraft guidance. There was no sign of a military connection in Edmund C. Berkeley's *Giant Brains or Machines that Think* of 1949. By now, authors such as Berkeley and Low were projecting a wide range of applications for the computers being developed in various academic and industrial departments. Most involved the machines simply taking over tasks now done by humans, generating fears of unemployment. The great advantage was the elimination of drudgery. *Popular Mechanics* reported their potential use for weather prediction and medical diagnoses.[36]

The computers of the 1950s still used valves and were necessarily huge and consumed vast amounts of power. Asimov published a number of stories imagining the United States governed by a huge computer named Multivac – the term derived from a real-world machine, UNIVAC. Multivac occupied a space equivalent to a whole city block and was built deep underground for security. Projected for the year 2008, it could specify the result of a presidential election by questioning a single individual to allow for human idiosyncracy.[37] Corporations such as IBM were built on the assumption that computers were large-scale machines that would be used only by government and big business. At this point, no one realized that the invention of the transistor would revolutionize the technology by making the computer available in the home or even in the pocket. Yet Asimov himself was already writing his robot stories in which a machine fitting inside an artificial human skull was imagined sophisticated enough to make it more or less autonomous. Because this involved a totally hypothetical 'positronic brain', it does not seem to have occurred to anyone – Asimov included – that in principle the suggestion made the whole idea of the huge computer potentially redundant.

Asimov's robots were depicted as more or less autonomous, though bound by the 'three laws of robotics' to protect human life. The possibility of public hostility to computers based on fears aroused by the Frankenstein story and Čapek's *RUR* was recognized in Berkeley's 1949 book. Low dismissed such concerns by noting that we could always just switch the machines off. This scenario was later dramatized to great effect in the 1968 movie of Arthur C. Clarke's *2001: A Space Odyssey*, when the spaceship's almost human computer HAL is disabled because it has begun to murder the crew. It was precisely to counter such fears that Asimov had introduced his three laws. Another counter to the 'dangerous computer' myth came in Robert A. Heinlein's 1966 story *The Moon is a Harsh Mistress*, where the machine running the lunar base achieves consciousness and aids the colonists throwing off oppressive rule from Earth.

By the 1960s, anticipations of the computer's future uses had begun to expand in directions we now recognize as perceptive of their wide potential. In 1967, the Electronic Computing Home Operator (ECHO) was predicted that would coordinate and control all household functions – although it would take up twenty square-feet of space in the basement. It would be a short step from this to Heinlein's vision of a computer capable of running a whole lunar colony. By the 1960s, many commentators were beginning to recognize the darker side of this potential. Thinking ahead to 1984, Dr M. V. Wilkes of Cambridge University warned that computers would allow states to monitor the activities of their citizens more closely, a concern also noted by Nigel Calder in 1969 when he included this as a 'Non-Standard Horror' to balance the boon conferred by the elimination of drudgery. This was in his *Technopolis*, where these more speculative concerns were added to the 'Standard Horrors' of science, including thalidomide, pesticides, fallout and pollution.[38]

There were hints at a more positive future. By the 1960s, more imaginative speculations were recognizing the potential for miniaturization, with computers predicted for the pocket or even in the wristwatch.[39] Wilkes saw both the positive and the negative potentials, noting the possible use of computers as the basis for a worldwide messaging service and other forms of global integration. Here at last was a serious anticipation of something like the internet, although it would be the 1980s before the transition to a more individualist cyberculture began to emerge to challenge the older fears of centralized control.[40]

6   Getting Around

For the technophiles, one of the most exciting prospects opening up was that of better transportation. Aviation created such a flurry of expectation that it deserves a chapter all to itself, but even the more mundane areas of terrestrial and marine travel offered opportunities for things to get faster and more convenient. Most of the new proposals focused on improvements to existing systems and in some cases were fairly rapidly achieved. More efficient engines would provide not only speed, but also economy, extending the availability of better transport to all. Streamlining cars and trains was presented as a way forward, although the improvements were limited and had their impact more through cosmetic attraction and advertising. The main focus was on motoring because it seemed to offer the prospect of individualized transportation (the same expectation drove some of the enthusiasm for aviation, with much less likelihood of fulfilment). Building on the revolution in manufacturing generated by the assembly-line system, the industry widened its appeal and its advertising to create the image of a world transformed for the better by the freedom to travel.

In fact, the availability of cars, coupled with their ever-faster speeds, was creating new problems. No one could ignore the congestion and pollution produced in city centres, while road accidents multiplied. Here was the foundation for a movement to reject the new developments altogether. Conservatives lamented the destruction of both the countryside and ancient city centres brought about by the need for more roads. Yet, as we have seen in previous chapters, there were more forward-looking architects and planners who relished the prospect of redrawing the map to create a framework for the new technology. Some also promoted improvements in public transport to head off the expansion of the roads. Streamlined trains, driven perhaps by electricity or some other new means of propulsion, were widely promoted.

Some proposals struck out in new directions and these enjoyed variable fates. There was a temptation to exaggerate the potential impact of innovations that might be genuinely useful, but only in certain circumstances. The enthusiasm of the visionaries often led them to ignore practical or social barriers that would impede wide application of their ideas. J. F. C. Fuller's contribution to the 'Today and Tomorrow' series saw transportation being revolutionized by the use of caterpillar tracks, eliminating the need to build roads in undeveloped areas.[1] Monorails were widely touted as the way forward for public transport, often imagined as being propelled by more or less the same means as aircraft. Again, the technology proved to have a limited application. Louis Brennan convinced many including H. G. Wells that his gyroscopic monorail would revolutionize rail transport, but the system never got beyond the prototype stage. In the 1950s, enthusiasts for the hovercraft echoed Fuller's claim that here was a breakthrough that would transform the global transportation network.

The same elements of enthusiasm and exaggeration can be seen in other areas. Tunnels were proposed to allow ground transportation to underpass waterways such as the English Channel. The Channel Tunnel was eventually built, but in the 1930s it was widely claimed that there would soon be tunnels underpassing whole oceans. Science fiction authors envisioned the complete replacement of motor transport by the 'moving ways' that Wells imagined for the megalopolis of *The Sleeper Awakes*. This might become necessary if the supply of oil dried up so drastically that governments would have to ban private vehicles. Moving ways have indeed gained limited application in airports and other areas where there is high-density pedestrian activity, but the notion that they could operate on a cross-country basis now seems ridiculous.

The automobile was the great success story of land-based transportation in the twentieth century, with huge consequences for how society was organized. It threated traditional rivals, especially the railways. In America, some sub-urban railways were in effect sabotaged by the activities of the automobile industry as it sought to expand the network of freeways. Coupled with the rise of aviation for long-distance transport, the railway system in the United States went into steep decline. In other countries, railways made something of a comeback as improved systems allowed them to provide inter-city transport at speeds sufficient to compete with aviation. The most successful technologies have been based on upgrading the existing rail network, not on the more exotic alternatives promoted by inventors.

Motoring

At the turn of the century, trains and steamships had been available for decades and had already revolutionized public transport. Motor cars were very new and still very experimental, but they had the potential to open a new age of individual transportation and were seized on by a few pioneers who strove to make them more practical. Not everyone was convinced and some were positively hostile, but the problems were soon overcome. By the 1920s, the mass production of automobiles had become well established in America and production was expanding elsewhere. The world was being transformed, although it would take time for all the consequences to become apparent.

One early enthusiast was Rudyard Kipling. In 1900, he bought a house, Bateman's, standing on the top of a hill that had 'used up' the previous owner's carriage horses. His motor car solved the problem, although he was assured that they would never catch on.[2] Kipling went on to write several stories about the hazards of motoring in an age of frequent breakdowns, few repair or refuelling facilities, and hostile policemen. At this point, there was a sharp divide between the few enthusiasts and a general public that was suspicious of the new machines. The Great War changed attitudes by making motor transport more reliable and more readily available. Henry Ford's assembly lines had already begun to mass-produce cars and European firms gradually caught up. By 1927, 10 million homes in America had a car, while in Britain car ownership rose from 132,000 in 1912 to over a million in 1930. Like Kipling, men (and they were mostly men at first) enjoyed the freedom of the open road. Driving became a symbol of masculinity. Aldous Huxley enjoyed motor touring on the Continent and in America despite his poor eyesight (and his later deprecation of Ford in *Brave New World*). Meanwhile, Le Corbusier enthused over the sense of humanity's power over nature generated by the excitement of speeding city traffic.[3]

Motoring became a key focus of interest in newspapers and magazines, especially those devoted to technical matters. It was a short step from reporting the latest developments to speculating about what was to come, and potential improvements were widely touted. Innovations in design and engineering, real and imaginary, proliferated. In the early 1920s, *Conquest* hailed the potential advantages of electric motors for cars and speculated about new fuels. Alcohol was a possibility if only the government would relax the regulations on its production. A reader who enquired about the possibility of hydrogen as a fuel was told that it diffused so rapidly that storage was a problem.[4]

Speculations continued into the 1930s as the technologies did indeed improve. *Armchair Science* wondered how long it would take for the weight of diesel engines to be reduced so they could be used in cars as well as in trucks. The magazine also raised more fundamental design issues, noting the potential advantages of a switch to rear-engined cars. The most radical proposal came in the form of Buckminster Fuller's Dymaxion car, which threw all conventions out of the window by having three wheels so it could turn on a circle (Figure 6.1). Unfortunately, a prototype was involved in an accident at the World's Fair in 1933, generating a wave of negative publicity.[5] In December 1942, *Practical Mechanics* featured a proposed three-wheeled,

Fig. 6.1. Buckminster Fuller's Dymaxion car, 1933. D. Bush,
The Streamlined Decade (1970), p. 106.

gas-turbine car on its front cover. By the late 1940s, the incipient nuclear industry also had its sights set on revolutionizing the automobile using small-scale reactors. Enthusiasts predicted a rapid move to clean, cheap cars that never needed to be refuelled. The warnings of experts that it would prove impossible to construct and shield reactors this small turned out to be closer to the mark.[6]

Some proposals moved to entirely different means of locomotion. In 1943, *Popular Mechanics* predicted a 'helicar', while in 1957 it hailed the possibility of a fan-driven car that could also take to the air.[7] At this point, motoring speculation merged with the hope that private aircraft would become available to all.

In addition to speculations about motive power, these proposals brought in two additional areas of automotive design: new materials and streamlining. The two areas were interconnected – aluminium and the new plastic materials were much better adapted than steel to producing the smooth lines required for streamlined bodywork. Science also played a role as the study of aerodynamics using wind-tunnels gave a better understanding of how air resistance reduced performance. This was really more important for the burgeoning aviation industry, but it had some relevance for motoring as speeds increased. Racing cars certainly benefitted from the introduction of streamlining and this added to the glamorous image created by designers and advertisers trying to convince the public that they were about to produce the 'car of the future'. At the start of the motoring era, the Futurist artist Fillipo Tomasi Marinetti had declared: 'A racing car is more beautiful than the winged victory of Samothace.'[8] Now that image of exciting modernity was being offered to the general public.

As early as 1913, *Scientific American* claimed that automobiles would eventually be designed with an aerodynamic teardrop shape. In 1922, a streamlined car produced by Rumpler in Germany was being promoted as the way forward, offering a 50 per cent reduction in fuel consumption. Three years later, A. M. Low predicted a streamlined car with 'wings' that served as aerials for radio communication.[9] In the 1930s, the streamlined look began to appear especially in American luxury cars and was seized on by advertisers seeking to create an image of modernity. Aluminium producers such as ALCOA and the Bohn corporation published striking advertisements depicting futuristic designs for streamlined cars, buses, trains and even agricultural machinery (see colour plate 3). Norman Bel Geddes also exploited the streamlined approach in his designs for projected cars and buses, pointing to Malcolm Campbell's record-breaking car, *Blue Bird*, as evidence that the aerodynamic profile really did increase speed and efficiency. In 1940, *Popular Mechanics* imagined cars of the near future that would be streamlined and air-conditioned – and made of synthetic materials. Henry Ford was photographed swinging a sledgehammer to illustrate the strength of a synthetic car body, although the outbreak of war blocked progress in this area.

Buses were less glamorous, but even here it was assumed that if they survived in an age of private motoring they would be streamlined for speed. Low thought that buses would disappear in towns, but predicted a role for fast, luxurious buses for long-distance travel. Various streamlined designs were planned in the 1930s and 1940s. In 1956, the Vibarti company in Italy proposed an aerodynamic bus built of plastic and powered by a gas turbine that would cruise at 125 miles per hour. Here, as with private cars, streamlining was assumed to go hand in hand with the move away from steel as the main construction material. In fact, the promise of plastic vehicle bodies turned out to be technically more difficult to realize than the enthusiasts had predicted and it was only later in the century that the move to synthetics began to take shape.[10]

The more extravagant visions may not have been realized, but there were huge improvements in propulsion and design which transformed the automobile from a subject of ridicule to a luxury item and finally into something available to all. The implications soon became obvious – towns and the countryside would have to be transformed to provide the infrastructure needed to cope with the increased traffic and the speeds involved. Some proposals centred on simply improving the road surface. Motor traffic demanded smooth paved roads to fulfil its potential and slowly the arterial roads began to be improved. There were suggestions for new road surfaces. At the start of the century, H. G. Wells imagined a world in which road construction had been revolutionized by a rubber-like substance he called Eadhamite. He was not the only author to imagine that rubber roads would allow for faster and quieter running. There was also a proposal to mould waste glass into specially designed blocks for road building. Low thought roads within cities would eventually be roofed over.[11]

Many of the traffic management techniques we take for granted today had to be worked out. One-way streets had been around for some time, but as the traffic in cities increased, police control at intersections became inadequate and traffic lights were introduced in the 1920s. Low predicted that soon traffic lights would be equipped with sensors to detect oncoming traffic, while *Popular Mechanics* thought the cars themselves would have a display indicating the state of the traffic lights ahead. It also imagined balloons carrying overhead cameras to relay information on traffic to a control centre.[12] But still the congestion increased and we saw in Chapter 4 how many town planners decided that the only way forward would be to separate motor and pedestrian traffic onto different levels, perhaps with a third, lower level for goods vehicles or trains. Air pollution was already becoming a problem in the 1920s and for all his enthusiasm Le Corbusier quoted newspaper headlines about the dying trees of the Paris boulevards.[13] Along with other radical planners, he wanted to rebuild city centres completely to harmonize the relationship between people, buildings and traffic. These more extravagant schemes have often foundered on public reluctance to destroy much-loved architectural features, although they have sometimes been realized in new cities or those rebuilt after war.

In the countryside, the spread of the arterial roads was relentless and enthusiasts called for measures to improve speed, paving the way for the motorways of today. By 1925, Low predicted that there would have to be separate lanes for traffic moving at different speeds. Looking ahead fifty years from 1933, he saw multi-lane highways with minimum speeds for traffic. Lockhart-Mummery agreed that there would have to be minimum speeds and suggested that the motorways might be built along the routes of abandoned railways.[14] The British were, in fact, well behind other countries in facing up to the challenge. Already in the 1920s, New York was building parkways to allow well-off citizens to pass through the blight of the suburbs to reach open country. Low noted that there were plans for a 100 mph New York-to-Boston freeway. The Germans began to plan their autobahn network in the 1920s, although building only began after the Nazis came to power. By 1935, the Italians had 500 kilometres of autostrada. The British had to wait until 1959 for their first motorway.[15]

Like the streamlined cars, these new roads were publicized as symbols of modernity and pointers to how the whole world would be transformed. They were the brainchild of engineers and technocrats who saw it as their job to bring about the transformation. When the German and Italian systems were introduced, they were ahead of their time in the sense that European traffic densities were not yet sufficient to demand them. To begin with, there was often serious effort devoted to achieving an aesthetically pleasing look that would harmonize with the countryside. Imaginative design was seen as a synthesis of modernizing technology and cultural sensitivity. Norman Bel Geddes's 'Futurama' exhibit at the 1939 New York World's Fair projected General Motors' preferred vision of the world in 1960 in which new cities were linked by superhighways. But already GM had begun to buy up and run down suburban electric rail systems to encourage the building of the roads their vehicles required. The result would be cities like Los Angeles, completely dependent on the automobile. More ruthless planners began to force through freeways that would destroy whole neighbourhoods.[16] The resulting tensions remain with us today.

The problems generated by urban traffic were not the only source of opposition to the motoring lobby, especially in Europe. Motoring provided opportunities to expand the suburbs and also improved access to the countryside and the seashore. Many welcomed these developments, but those with more conservative attitudes lamented the destruction of traditional landscapes and ways of life. In Britain, writers such as C. E. M. Joad and Clough Williams-Ellis pleaded for the countryside to be protected from the onslaught of the motorists.[17] Efforts to make the new motorways blend in with the countryside through which they passed were a belated response to these fears, but for the opponents this would always be too little and too late. They could also point to the carnage resulting from traffic accidents – in 1930 there were 7,000 deaths on Britain's roads (over twice the current figure despite the huge increase in car ownership). Paradoxically, the new designs for superhighways helped to make driving safer as well as faster, even as they helped the motoring lobby to impose its views on modern culture.

Moving Ways

The expansion of road traffic posed a threat to established alternatives, especially the railways. The latter hit back with some success as we shall see in the next section, but there were a few visionaries who thought that all existing systems based on moving vehicles might eventually be replaced. In *The Sleeper Awakes*, H. G. Wells had portrayed a huge megalopolis in which vast numbers of people were transported around by the 'moving ways' – continuous belts that could be accessed by simply stepping on and off. The technology was certainly feasible – during the inter-war years people became used to the escalator or moving staircase in train stations, etc. and to adapt the system to horizontal movement was obviously possible. Many visions of the future predicted cities in which these moving ways would replace road transport altogether. A few science fiction authors went further and imagined the technology being applied on a massive scale for cross-country movements. In part, these ideas were a response to the growing problem of congestion on the roads, but there was also a concern that if the supply of oil ran out (a consistent fear as we shall see in Chapter 10) a form of transportation that could be powered by other sources would be necessary.

Wells's novel, first published as *When the Sleeper Wakes* in 1899, imagined humanity concentrated in giant mega-cities in which the transportation was by 'moving ways' powered by the electricity generated by 'windvanes' on the city roofs. The technology to create a moving sidewalk horizontally or an escalator to change levels soon became available on a small scale and was used at exhibitions and increasingly for access to underground railways. To imagine this rather crude mechanical equipment scaled up so that it could become the main way of getting people around a city seemed fairly straightforward. As traffic congestion and pollution in the big cities increased, several of the writers who followed Wells in trying to predict the future thought this might be the only solution. In 1925, Low suggested moving sidewalks in cities, while a decade later, Lockhart-Mummery imagined two-speed moving walkways in the London of 2456.[18] In fact, the expansion of underground railways and better traffic management above ground managed to cope with the expansion of the urban population. Moving walkways and escalators did become part of everyday life, but only in limited circumstances.

Speculation in this direction re-emerged in the 1940s among the rising stars of science fiction writing. Robert A. Heinlein's story 'The Roads Must Roll' imagined a world in which inter-city transport takes place by moving roads, with strips running at various speeds so passengers could step onto the outer strip and then move up to a central area with seats and even restaurants moving at great speed. The technology has been introduced because the depletion of the Earth's oil resources has made both private and goods transportation by vehicles impossible. The roads now run on solar power.[19]

Shortage of resources produced in part by an expanding population was also central to the imagined world of Isaac Asimov's *The Caves of Steel* of 1954, where again the cities use moving expressways as the main means of transportation. The old roads still exist buried underneath the city, but are only used by the police and emergency services. Asimov still thought the technology would be employed in the future when he made his predictions after visiting the World's Fair in 1964. Arthur C. Clarke went a step further in *The City and the Stars*, where the city Diaspar has moving ways driven by a material that flows like a liquid in one dimension only. Even so, when the hero finally escapes the city to travel across the world, transportation is by an underground railway system that Clarke admits would be familiar to anyone from the distant past.

Rescuing the Railways

The railways were a technology of the previous century, but the systems were well established all over the world and resisted the onslaught of the motor vehicle. In the cities, underground and elevated systems proliferated, especially now that electrification could be easily applied over the relatively short distances involved. Cross-country was a different matter. Here, there were many suggestions for ways of improving the speed and efficiency of the existing lines. Electrification was more challenging on this scale, so many hopes were pinned on the development of diesel and diesel-electric propulsion. Streamlining could be applied to trains as easily as to any other form of rapid transportation, although its effects were more cosmetic than practical. There were also much bolder proposals centred on the invention of various kinds of monorails, often using designs inspired by the latest developments in aviation. These were flagged in the popular press with monotonous regularity, usually accompanied by predictions of vastly increased speeds. In the end, most of these ideas failed to win commercial approval. The most radical, Louis Brennan's gyroscopic monorail, offers a classic case study of a technology which was initially hailed as revolutionary, but was unable to find a place in the real world.

The moving ways imagined by Wells and others failed to gain wide acceptance, in part because installing a mechanical system on every city street would create huge problems of integration and coordination. Underground railways offered a more practical system – they had been around in some big cities even in the days of steam, but were much more convenient and efficient now they could be electrically powered. The people of London and other great cities soon became proud of their underground systems and looked forward to extensions and improvements. There were alternatives that could be applied in circumstances where there was high traffic density over a well-defined route. Continuous railways running at slow speeds so they could be accessed by pedestrians without stopping were proposed and actually used, for instance at the British Empire Exhibition of 1924. This was, in effect, an equivalent of the moving way system and could really only work in the same environment. Even so, it was hailed as 'The Railway of Tomorrow'.[20]

On a wider scale, there were plenty of proposals for modernizing the existing rail networks. Some of the most forward-looking related to freight transportation, inevitably the least glamorous aspect of the system and hence the least likely to be widely discussed in the popular press. But experts and the technical publications were well aware of the need for modernization, with the most important proposals centring on containerization. In 1919, engineer A. W. Gettie proposed a 'Clearing House' in central London where freight containers would be mechanically transferred from long-distance rail to underground rail or conventional road transport for local distribution. He claimed it would save the country a million pounds a day, in effect covering the financial costs of the Great War. He also suggested a similar scheme at the London docks. His ideas received some support, but also attracted the opposition of vested interests in the industry. As Sir William White, a former naval constructor, said to Gettie: 'You have proved the railway managers to be a pack of fools, and they will never forgive you.'[21] In the end, containerization proved to be the way forward, although the case was still having to be made in the 1960s. In *The World in 1984*, Camille Martin of France's SNCF insisted that containers for goods were essential for efficiency, a point the British would be among the last to accept.[22]

In the more publicity-conscious world of inter-city passenger transport, the emphasis was increasingly on speed. Here, the most visible challenger was the newly emerging aviation industry and the railways needed to counter the impression that they would never advance beyond the achievements of the previous century. The airliners of the 1920s actually had speeds scarcely faster than the best express trains, and trains had the advantage that they could start and finish in city centres. But planes were certainly going to go faster in the future and the railways had to show that they could improve their performance. There were also questions of efficiency and pollution, especially with the traditional steam engine.

Steam was yesterday's technology, but it might be possible to get at least some improvements if serious technical research was undertaken. The possibilities were still being actively debated in the 1930s. In Britain, *Meccano Magazine* promised the country's juvenile trainspotters that steam could do better if higher pressures were employed. *Armchair Science* claimed that soon there would be trains running at a hundred miles an hour, although it conceded that this was approaching the limit of steam's efficiency.

J. N. Leonard's *Tools of Tomorrow* cautioned that while higher speeds were possible, they may not be economic. He insisted that streamlining would be of little real value.[23]

If steam was approaching its limits, the race was on to develop alternatives that would offer better performance – and to present them as just as 'modern' as the flashy new cars and aircraft. The two most promising forms of motive power that could be adapted to the existing systems were diesel engines and electrification. The latter was probably going to be more efficient, but required major additions to the infrastructure of the tracks. As early as 1905, *Popular Mechanics* told its readers of a diesel-electric design being developed for the Southern Pacific Railway. By the 1930s, diesels were starting to appear more widely in the United States, while Ritchie Calder told British readers to expect them soon.[24] It would be well after the war that the British decided very abruptly to phase out steam in favour of diesels. They did have an extensive debate on the merits of electrification in the 1930s. *Armchair Science* told its readers about the recommendations of the Weir Committee which projected a national scheme that would take twenty years to realize at a cost of £261 million (plus another £45 million for suburban lines and £80 million for new power stations). The problem was that while diesels could simply replace steam engines, whole lines had to be electrified before the system could work at all. Vernon Sommerfeld cautioned that it would only work for lines carrying heavy traffic. Everyone agreed that it would be possible – although in fact nothing was done and some British routes await electrification to this day.[25]

One technical development that could be applied to any form of motive power was streamlining. As Leonard pointed out, it was of little real value unless speeds could be pushed to 100 miles per hour and beyond, although there would be some savings in efficiency. Streamlining was as much symbolic as practical – its benefits were trivial, but it helped to persuade the public that the railways were capable of moving with the times. The mania for streamlining that influenced the automobile designers also spread to the railways. The diesels introduced in America were publicized to create the impression that they pointed to the world of tomorrow (colour plate 3). The British could only look on enviously as the Americans, Germans and Italians went ahead with their futuristic designs. As Sommerfeld told them in 1935, the diesel Flying Hamburger was now the fastest train in the world.[26] Streamlining could be applied even to steam engines, although any benefits it conferred were liable to be offset because of the cladding obstructing access for maintenance. The author remembers Sir Nigel Gresley's streamlined A6 locomotives still at work in the 1950s, some of which are preserved today by enthusiasts convinced that the age of steam was ended too abruptly in the following decade.

Many predicted the eventual demise of the rail industry, and in some countries there was certainly a steep decline in its use. In America, the combination of an aggressive promotion of the automobile industry and the rise of aviation for long-distance transport led to the virtual elimination of the cross-country rail network. Elsewhere, the traditional system made a comeback thanks to the development of new technologies that at last allowed trains to reach the speeds predicted for some more exotic alternatives earlier in the century. Electrification and the development of dedicated high-speed tracks allowed intercity trains to reach speeds of 200 miles an hour or more, at which level they can compete with airlines over shorter routes. France especially saw electrification as the way forward. In June 1941, in the very same issue that detailed the terms of the Armistice with the invading Germans, the magazine *L'Illustration* included an account of plans to extend electrified lines to link Paris with Marseilles and Nice, with speeds up to 140 kilometres per hour. Streamlined steam locomotives were dismissed as a dead end. By the early 1960s, young enthusiasts were told of hopes to achieve speeds of 330 kilometres per hour. The French TGV system was not inaugurated until 1981, but the Japanese 'Bullet Train' was introduced as early as 1964. They have shown how improvements in an existing technology can, in the end, do better than some of the innovations that were hailed as potential replacements.[27]

Monorails

Sommerfeld noted that some of the most imaginative designs for future railways involved abandoning the existing track system altogether. For many enthusiasts, monorails offered the best way forward. The most radical design, proposed at the very beginning of the century, offered greater speeds long before anyone associated this with the streamlined form. But the Brennan monorail depended on a gyroscope to keep it stable and fears for its reliability meant that its moment of glory was brief. During the interwar years, there were numerous proposals for less radically balanced systems, some with the rail above, some below. The growing enthusiasm for aviation prompted designs that mimicked the increasingly streamlined form of the latest aircraft and even copied their means of propulsion – the propeller. There was also a constant stream of radical proposals for reducing the friction between the vehicle and the rail by electromagnetic and other futuristic effects. Only a few of these ideas were ever put into practice and like the moving way and the continuous railway, the monorail remained a niche product lacking the potential to revolutionize the whole industry.

Louis Brennan was an Australian engineer who invented an improved torpedo and was given an appointment at the Royal Navy's dockyard at Chatham. Here, he developed his gyroscopic monorail, based on the unique stabilizing properties of the spinning gyroscope which allowed a railway carriage to balance on a single rail. Any tendency to tilt would be automatically corrected by the gyroscope, so the carriage could run on a single narrow rail, or even a cable. He was given a War Office grant to develop the idea, on the assumption that a single track would be quicker to lay under wartime conditions. It would also be cheaper to lay – £1,000 per mile was projected as opposed to thirty times that amount for conventional track. Brennan built a working model and then a fully functional prototype, filing a patent in 1903 and then developing the system over the following decade. It was demonstrated at a White City exhibition in 1910. Brennan's working design involved cars driven by petrol engines, each with two one-and-a-half-ton gyroscopes enclosed in evacuated vessels and carried by specially designed bearings allowing them to spin for an hour and a half even if the engine cut out. Realizing that strong gusts from side winds might be too much for the gyroscopes to cope with, he designed a system that would use compressed air to apply a greater correction when needed. Speeds of up to 130 miles per hour were projected.

There was huge public interest in Brennan's project and it was widely anticipated that it would revolutionize the whole transportation network. Not only would the speed of travel be increased, but lines could be laid more widely across the countryside. It was suggested that bridges across rivers etc. could be replaced by single cables, and there were hints of a project to bridge the English Channel with a cable supported by pylons, allowing the journey from London to Paris to be completed in a couple of hours. The *Harmsworth Popular Science* serial included a substantial account of Brennan's ideas along with illustrations and photographs of the prototype, as did *Popular Mechanics* in America (where the cheapness of track-laying would have been even more beneficial). Both printed artist's impressions of the monorail traversing a deep river gorge on its single cable – the cars were depicted as two- or three-decked, but looked otherwise like conventional railway carriages.[28] There was no sign of streamlining at this point, despite the high speeds projected. In his *The War in the Air* of 1908, H. G. Wells imagined a world where surface transportation had already been revolutionized by Brennan's system. He included an image of the monorail crossing the Channel on huge pylons with ships navigating below (Figure 6.2).

War did indeed break out in 1914 and Wells's predictions about the role of air power were partly vindicated. But Brennan's invention seems to have been largely forgotten as the war progressed – it was certainly not taken up by the

" PRESENTLY THE ENGLISH CHANNEL WAS BRIDGED."

[To face p. 14.

Fig. 6.2. The Brennan gyroscopic monorail crossing the English Channel.
Note also the airships. From H. G. Wells's *The War in the Air* (1908),
facing p. 14.

military authorities. There were a few brief references to it in post-war publications, but mostly in conjunction with the new generation of less radical monorail proposals. One exception is a 1933 article in *Meccano Magazine*, accompanied by a front cover repeating the vertiginous image of a car suspended on a single cable across a deep gorge.[29] The magazine predicted that the system would indeed revolutionize transport, although it also promoted less radical designs.

The whole episode presents a classic illustration of an invention that is greeted with huge enthusiasm by the press but which fails to capture the support of the industry concerned. There would be substantial vested interests operating against Brennan's proposals, but it is significant that one of the few post-war references to the system mentions 'reliability' problems as the source of the industry's reluctance to take it seriously.[30] Despite the demonstration at White City, the images of carriages crossing huge gulfs balanced on a single cable may have been counterproductive at the psychological level. It would be hard to imagine the ordinary citizen feeling safe in these circumstances, whatever Brennan's assurances about the reliability of his gyroscopes. The metaphorical rise and fall of the Brennan monorail forces us to invoke a multiplicity of factors when we evaluate the forces governing the actual course of progress in any area of technology.

In its account of Brennan's invention, the *Harmsworth Popular Science* provided a photograph of a rival German system in which the car was suspended from an overhead rail. Similar projects proliferated in the post-war era, offering ever-greater speeds as railways faced the threat generated by aviation. The greatest enthusiasm was in Europe – America seems to have focused more on simply improving the existing networks. Nevertheless, the American public was kept informed. In 1928, *Popular Mechanics* featured a design by the Schutze-Lanz airship company (a rival of Zeppelin) for an overhead monorail system that was intended to run at 200 miles per hour on the Berlin–Ruhr route. This was streamlined and driven by an aircraft-style propeller. Three years later, the magazine featured another German design intended to run at 150 miles per hour and promised that something similar would soon be running in New York. In 1940, it depicted a system riding above the single rail, again streamlined and propeller-driven, and with an aircraft-style tailfin for stability. Walter Kaempffert described a similar system.[31] In the 1930s, the British tried their hand with the 'Railplane' system designed by Glasgow engineer George Bennie, using a streamlined car slung beneath the rail, again driven by airscrews. A prototype was built at Milnagarvie near Glasgow and featured widely in the British press. After the war, there was a plan to use the system in central London and in the Egyptian desert, but nothing came of these projects. In 1962, juvenile French readers were told that streamlined monorails would become commonplace in European cities, and several systems were in fact built.[32]

Reduced friction was one advantage offered by a monorail system, and some more adventurous engineers realized that this benefit could be enhanced if the car did not actually touch the line. In 1932, *Meccano Magazine* told its juvenile readers about a design proposed by German engineer H. F. Kutschbach, in which speeds of up to 600 miles per hour were to be achieved. Building on French experiments done before the war to raise the car above the lines by electromagnetic action, this scheme imagined streamlined cars projected through rings which kept them on course by the same force.[33] Such a system was depicted for terrestrial transport in one of Robert A. Heinlein's early science fiction novels, *Starman Jones*. The idea of using electromagnetic effects to lift the car above the rail continues to attract the attention of the more adventurous engineers.

An alternative approach was developed in the 1960s by Christopher Cockerell, the inventor of the hovercraft (discussed later in this chapter because its main application turned out to be for ferries). Cockerell applied the same principle of using compressed air to 'float' the car half an inch above the single rail. Speeds of 400 miles per hour were promised for the 'hovertrain'. In America, *Popular Mechanics* hailed the invention, while French readers were told that the principle would revolutionize not only trains, but also cars and ocean liners. In the end, none of these applications was realized.[34]

Although some monorails were built, the technology has never replaced the conventional railways. The hovertrain principle was considered by French engineers planning their TGV network, but rejected in favour of the traditional *chemin de fer*. In the end, it was the older technology, vastly upgraded by better tracks and electrification, that allowed the railways to implement services at speeds of up to 200 miles per hour on a commercial basis. For all the publicity accorded to the monorail pioneers, it was engineers looking for ways of improving the existing technology who allowed it to compete with aircraft, at least on short-haul routes.

Tunnels

There remained the problem of the stretches of water blocking land communication. The English Channel was the most obvious example of a barrier that lay across a route carrying a high density of traffic. The possibility of tunnelling under the Channel had been raised in the nineteenth century, and in the twentieth it began to seem within the bounds of technical feasibility. Tunnels were built to carry motor traffic under rivers, but no one thought that this would be appropriate for a tunnel that was thirty miles long, so the proposals were invariably based on the idea of running trains directly between Britain and

France. Much excitement was generated by a new proposal in the late 1920s, although the plan was never taken up and the project was not revived until the 1980s. Plans were also mooted to tunnel under other stretches of water, including the Straits of Gibraltar. Science fiction writers had a field day imagining even more extensive projects such as tunnels beneath the sea. These ideas seemed outlandish at the time, but they were occasionally revived later in the century by the advocates of systems involving trains moving through a tube evacuated of air.

There were occasional references to plans for a Channel Tunnel in the popular press during the early decades of the century.[35] The debate really got underway following the publication in 1928 of a book by engineer William Collard with the conventional title *Proposed London and Paris Railway*. His design called for a tunnel carrying a railway built on a seven-foot gauge (hence separate from the existing rail network). Streamlined trains would run at sixty miles per hour underground and twice that speed outside the tunnel, allowing Paris to be reached from London in less than three hours. Detailed plans were worked out and the total cost was estimated to be £189 million. Collard was aware of earlier fears that a tunnel could be used by an invasion force, but insisted that it could easily be blocked if necessary. A draft report by the Marquess of Londonderry was included in the book, noting that an enemy would have to seize both ends of the tunnel in order to transport an invading force.

Collard's scheme was widely discussed in the technical literature and reached the public through articles in the popular press. *Armchair Science* carried an article in 1930, but had already been pre-empted by the juvenile *Meccano Magazine* in the previous year. Unfortunately for Collard, this juvenile magazine subsequently published a spoof story about a mad Professor Barmidotti planning to bridge the Channel with rockets. Children were also informed about the project in I. O. Evans's *The World of Tomorrow*, which referred to plans to blow up the tunnel in the case of invasion.[36] The scheme was never taken up and public interest waned as the international situation became more tense in the later 1930s.

There were also plans to use tunnels as links across other sea barriers. In 1935, the French, aware of the need to link with their North African territories, proposed a tunnel under the Straits of Gibraltar. This was to be a 'floating tunnel': soft material from the seabed was to be cleared by explosions and then a rigid tube would be laid on top of the bedrock to carry the railway track.[37]

These schemes were just about within the bounds of feasibility, but the writers of science fiction were not bound by such limits. They imagined tunnels being built of vast lengths, including under the Atlantic Ocean. Such ideas may have seemed bizarre, but trains could travel several times as fast as the latest ocean liners, so in principle significant time could be lopped off the transatlantic journey. Developments in commercial aviation soon began to undermine the credibility of these speculations, but they retained some plausibility in the early days when aircraft speeds were still limited. Airships might cross the Atlantic, but they were even slower.

Michel Verne, son of Jules Verne, had published a story about a transatlantic tunnel in 1888, translated into English in 1895 (and mistakenly attributed to the father). Another suggestion came in a successful novel published in 1913 by the German writer Bernhard Kellermann, well before the era of commercial air transport. Hugo Gernsback's 'Ralph 124C 41+' also mentioned a tunnel linking New York to France. Kellermann's book was used as the basis for a film *Der Tunnel* in 1933, remade for English-speaking audiences in 1935.[38] By this time, transatlantic flight had become a reality, although not yet on a commercial basis. The movie imagined the project taking twenty years to complete (by which time aircraft would be flying the Atlantic at far greater speeds than the trains in the tunnel).

The idea of long-distance tunnels was revived in the post-war era by scientists planning entirely new systems that would travel even faster than the commercial jets then becoming available. In 1950, Irving Langmuir proposed to link New York and San Francisco by a tunnel that would be evacuated of air, allowing trains to travel through it at speeds of up to 5,000 miles per hour.[39] Projects based on the same principle have never materialized, although less ambitious projects were eventually successful. Hopes were high for the construction of a tunnel under the English Channel in the 1960s, with *New Scientist* warning that if successful there would be a massive build-up in the population of South-East England. The French were also keen, although they also considered a rival scheme based on a bridge.[40] The Channel Tunnel did eventually get built, opening in 1994. Once again, the more restricted version of an idea eventually gained enough support to be realized in practice, while the wilder speculations have remained in the sphere of fiction or the more speculative forms of journalism.

Shipping

At the start of the century, ships remained the only way to cross the oceans. Steamships had revolutionized the industry during the later nineteenth century (although sail remained important for low-cost non-perishable cargoes). The ocean liners became the most glamorous means of long-distance transport for the well-to-do, less so for the poor, who had to travel steerage. Despite tragedies such as the sinking of the *Titanic* in 1912, the steamship companies could still appeal to the upper classes by offering ever-greater luxury and better speeds. But there was a limit to how fast an ocean liner could be driven economically, and the advent of aviation would ultimately destroy the shipping companies' hold on the transoceanic routes. As with the railways, but with less success in the long run, there were proposals for new technologies that might allow ships to move at ever-greater speeds. A few were plausible in certain circumstances: the hydrofoil and the hovercraft worked well for small vessels on sheltered waters, but plans for jet-propelled ocean liners and the like came to nothing. The liner was ultimately to be transformed into the cruise ship, a development already foreseen in the 1930s.

It was the transformation of freight traffic that saved the shipping industry, again a move prefigured in plans from the start of the century. There were some dissidents: as late as 1935, J. N. Leonard suggested that freighters had reached the limit of economic size, so no further developments could be expected.[41] In fact, the eventual move to containerization allowed ever bigger ships to be built and loaded efficiently: cargoes increased dramatically even as passengers switched to the airlines in the 1960s. These developments were seldom featured in the popular literature because they were seen as being of little interest to anyone outside the industry. Gattie's 1919 plan to revolutionize the freight distribution system in London did receive some attention in parliament and the press. It included a mechanized system for the Albert Docks that depended on containerization.[42] The proposed new wharf was never built, and when the new container ports were developed much later in the century, they were usually in locations outside the established trading cities.

Gattie also proposed to restructure the country's canal system, and here he was less out of touch with contemporary opinion. The early phase of Britain's Industrial Revolution had been based on canals for the transport of heavy goods and the system was still seen as viable. But the complex network of canals constructed to different specifications that had spanned the British landscape was in dire need of rationalization. A Royal Commission in 1909 noted the major developments in canal building on the Continent, where great rivers such as the Rhine had always served as a network for shipping. It argued that if the British network could be upgraded to similar specifications, the benefit to the country's industry would be considerable. Planned developments in the American canal system were also noted enviously in 1925, while ten years later there was a proposal for a canal network that would operate without locks to link the major industrial centres.[43] The British canals were never revitalized, but developments elsewhere in the world showed that in appropriate circumstances an old technology could be modernized to stave off competition from newer systems. For long-distance transport of heavy non-perishable goods, even the canals still had a role to play.

Through the early twentieth century, public interest focused on the great ocean liners and on warships. Both went through a series of innovations in design and motive power intended to make them bigger and faster. Until the advent of transoceanic aviation in the late 1930s, the liners were the only means of intercontinental travel and for first-class passengers at least they provided a level of luxury equivalent to what could be enjoyed at home. There was intense competition between the shipping lines to offer the fastest service, especially across the Atlantic. Various alternatives to the traditional steam engine were tried out in the hope of improving performance, along with other improvements such as streamlining. Even so, some perceptive commentators realized that the ships' days were numbered. In 1935, J. N. Leonard argued that although faster ocean liners were possible, they would be uneconomic and would be introduced solely to gain prestige. In the same year, Vernon Sommerfeld suggested that the liners were becoming little more than floating hotels for the well-to-do, a prediction vindicated by the emergence of the modern cruise ship.[44]

When the British launched the first all-big-gun battleship, HMS *Dreadnaught*, in 1906, she was powered by steam turbines, a new feature on a ship of this size. The steam was still generated by burning coal, however, although soon there would be a move to oil firing. The great advantage of oil for warships and ocean liners was the convenience of fuelling, a notoriously dirty business with coal. Turbo-electric power was also offered as a more flexible form of power. This improvement was suggested in response to the rise of another alternative, the diesel engine, also touted as the motive power of the 'ship of the future'. Already in 1907, *Popular Mechanics* claimed that the American and British navies were looking to the internal combustion engine.

A few years later, the *Harmsworth Popular Science* serial predicted the rise of diesel power, suggesting that it might paradoxically lead to a revival of sail. This was on the grounds that diesels made ideal auxiliary power plants for use when there was no wind (a point exploited by modern yachts). Shortly after the war, there was an attempt to modernize this combination in the form of the Flettner rotor ship, which used giant rotating cylinders to generate thrust at right angles to the wind from the Magnus effect. The inventor, Anton Flettner, pointed out that the system was more efficient and more flexible than sail. The ship needed power to drive the rotors, which were thus seen as a way of reducing fuel consumption when the wind was favourable. The rotor ship made a transatlantic voyage in 1926, but despite some enthusiastic speculation was never seen as economically feasible.[45]

By the 1930s, most warships and ocean liners had switched to oil power, a move lamented by Captain Bernard Acworth of the Royal Navy, who pointed out that Britain was completely dependent on foreign sources of oil, while she had plenty of coal reserves (this was long before the discovery of oil in the North Sea). Acworth was a noted conservative – we shall encounter him as an opponent of aviation in the next chapter – but his point would be vindicated in World War II when the German U-boats threatened to starve Britain of oil and other raw materials. One suggestion that would have pleased Acworth was the proposal to use pulverized coal as an alternative to oil.[46]

In the end, it was the diesel that triumphed. There were nagging fears about the eventual exhaustion of crude oil supplies (discussed in Chapter 10), but in the short term at least these were put aside. The great freighters and cruise ships of today are mostly driven by oil, although for a brief period in the 1950s and 1960s it seemed as though nuclear power might offer a new alternative. Given the size of reactors and the problem of shielding, ships were really the only form of transportation for which nuclear power could become a reality. Nuclear power allowed submarines to cruise for months underwater and was seized on by the world's navies in the Cold War. The first nuclear submarine, the USS *Nautilus*, was launched in 1955 and the first Russian boat three years later. The Americans and the French also have nuclear-powered aircraft carriers. But efforts to convert the merchant marine to nuclear power came to nothing. The United States built an experimental ship, the NS *Savannah*, in 1962, but neither she nor rivals built by other countries were commercially viable. There were also fears that collisions or other accidents would allow radiation to escape within a major port or busy seaway.[47]

Motive power was not the only focus of attention. Designers were well aware of the need to ensure a smooth motion of the hull through the water, and during the 'streamlined age' of the interwar years there were many plans for futuristic ocean liners, most of which were never built. The streamlined effect was largely cosmetic, although Norman Bel Geddes insisted that his proposal would cut a day off the transatlantic voyage in bad weather.[48] He planned to incorporate the lifeboats inside the streamlined shell. German plans went even further, linking streamlining to innovative new means of propulsion. A 1928 plan featured a system for sucking water into the hull at the front and expelling it aft, while in 1946 it was proposed to use jet engines (presumably turbines) attached to the hull beneath the water. A. M. Low thought the liner of the future would have a landing strip for aircraft so passengers could get ashore more quickly.[49]

For smaller vessels, there were several new ideas which enjoyed some commercial success. One was the hydrofoil, in which the bulk of the hull is lifted out of the water on skies to reduce drag. In 1928, *Popular Mechanics* featured a German design for a hydroplane that would carry 200 passengers across the Atlantic at twice the speed of the fastest ocean liner. This was never built, but by 1965 the magazine could report that hydrofoils were in regular use on lakes and in coastal waters, and was still hoping that they had a future on the transatlantic route. In fact, the system has limited ability to cope with rough seas, making voyages on the open ocean impractical.

The same problem has also bedevilled another invention, the hovercraft. Various suggestions for lifting a vessel above the surface on a cushion of air had been made earlier in the century, but the first successful design was produced by Christopher Cockerell in the late 1950s, with commercial production beginning in 1962. In Nigel Calder's book of predictions for 1984, Cockerell foresaw a great future for his system if only financial and legal problems could be overcome.[50] The hovercraft could move smoothly between land and water and could cope with modest irregularities and waves. The legal problems Cockerell recognized arose from the fact that the machines could hardly traverse freely across countryside under private ownership. Hovercraft did indeed come into service mainly as car ferries and for search-and-rescue craft. They also have military applications. But their use as ferries has declined in part because, like the hydrofoil, they cannot cope with really rough seas.

The fate of the hydrofoil and the hovercraft parallels that of many of the other bright ideas for new transport systems. Some seem so bizarre it is hard to imagine them ever being taken seriously, while others – hailed at first as having the potential to revolutionize our lives – have found uses only in specialized niche applications. Hydrofoils do operate successfully in sheltered waters, but the world's ocean traffic is still carried in conventional ships which may be larger and more sophisticated than those of earlier years, but are still powered by an old technology, the diesel engine. Monorails and moving ways have a role in land transport, but mostly in urban settings – rapid cross-country services are still based on upgraded versions of the traditional railways. All too often, practical and commercial factors limit the applicability of an idea that the inventor or enthusiast sees as revolutionary. Yet, every now and again, a new invention does revolutionize our whole way of life. The motoring pioneers were ridiculed, but lived to see their brainchild transform the way in which we travel so dramatically that it has shaped the development of society as a whole. There was another new mode of transportation in the early twentieth century that had the same potential: aviation.

7   Taking to the Air

Of all the new technologies developed in the early twentieth century, none was seen as more exciting by the general public than aviation. Once the initial skepticism had been overcome, there was widespread enthusiasm, driven in part by the expectation that eventually the ability to travel by air would become available to all. Newspaper proprietors offered substantial prizes encouraging pioneers to fly further or faster than anyone had before. They ensured that the attempts would be widely publicized and as a result huge crowds turned out to welcome figures such as Louis Blériot and Charles Lindbergh when they succeeded in breaking another barrier. Flying displays also became popular. Some enthusiasts claimed that aviation would unite the peoples of the world, although (as with radio) there were critics who doubted the ability of any form of communication to achieve this goal. As late as the 1920s, there were still skeptics who scoffed at the whole idea of a commercially viable aviation industry. Rapid technical developments soon allowed their arguments to be discounted.

There was, of course, a darker side. From the start, it was obvious that aviation would have military potentials and the Great War showed just what these might entail. The British in particular had seen this coming because it was clear that the Royal Navy could no longer offer a reliable defence of their island nation. The effects of bombing were already anticipated in novels such as Wells's *The War in the Air* of 1908 and within a few years German attacks on London anticipated the far greater horrors that would be perpetrated in later conflicts. Politicians and newspaper publishers called for stronger air forces as the only means of deterrence. Here we see at its sharpest the division between those who hailed and those who feared the new technologies, and in this case the two visions competed on an almost daily basis in the newspapers of the time. The same paper might print a description of some new speed or distance record written by a correspondent who was an enthusiastic proponent of civilian aviation alongside the warnings of a politician or military figure concerned about the nation's vulnerability to attack. Specialist books and magazines propounded their own interpretations, but the ordinary newspaper reader (or radio listener) was left to make what sense they could of the competing visions offered to them.

The debate over the benefits and risks of aviation raises a major issue addressed by historian David Edgerton.[1] He notes that most accounts have been written by enthusiasts who assume that aviation is an inherently civilian affair and that its natural development has been distorted by the demands of the military. On this model, it was the developments pointing the way toward commercial airlines that were the real driving force of progress (reflected in the enthusiastic endorsements of aviation correspondents). Wars may have spurred progress, but this would only happen during occasional episodes of dire necessity. Edgerton argues that the aviation industry was from the start driven mostly by military considerations. Much of the technical progress made was initiated by government defence funding and only later made available for the manufacture of civilian machines. If he is right, then the pessimists of the time had a better interpretation of what was going on than the enthusiasts. Yet there can be no doubt that ordinary people were attracted to the prospect of mass civilian aviation and the crowds who turned up at air displays were to some extent blind to the actual use to which the machines they saw might be put. They could still be awed by the power of machines that they knew might soon be used to kill them in their beds. They hoped to travel in airliners even though the most impressive displays were by military planes. This kind of doublethink played a crucial role in efforts to predict future technical developments and their implications.

Warnings of the harmful applications of aviation dovetailed with those predicting other more directly warlike technologies such as poison gases and the atomic bomb. For this reason, the writings of the pessimists who focused primarily on the use of aircraft as weapons will be postponed until Chapter 9 below. But this division is made in the full knowledge that it is to some extent artificial as far as the actual technical developments are concerned. Edgerton is surely right to insist that we cannot treat the rise of the aviation industry as primarily a civilian concern. Yet the division between the enthusiasts and the pessimists has some validity when we consider the public perception of what was going on rather than the actual research that drove the production of more advanced flying machines. The rhetoric extolling the excitement of new developments and their potential to revolutionize everyday life for the better was certainly real, however one-sided it may appear in hindsight.

Many of the advantages promised by aviation were equally attractive for both civilian and military purposes. Aircraft could traverse over any obstacle on the surface, including stretches of water. What was needed to confirm their advantage over ships and trains was greater speed and greater range. The key question was how this was to be achieved. Several different technologies became available and for some time it was by no means clear which would prevail. The first half of the century saw intense competition between rival projects, each with supporters convinced their option was the best way forward. Land-based aeroplanes, flying boats and airships all had their advocates. Each had advantages and disadvantages which designers and engineers strove to exploit or eliminate. Speed was a major issue, as was the question of how to make airports easily accessible from city centres. Some enthusiasts longed for personal aircraft to replace the motor car. Crossing the oceans was at first a major problem for land-based planes and some bizarre ideas were floated to deal with this. Eventually, straightforward technical advances produced aeroplanes that could fly at high speeds for thousands of miles non-stop, finally ushering in the much-heralded age of mass air transport. The last stages in this progress were, however, driven by the design of bombers during World War II.

Filling in the details of these proposals provides a fascinating window into the complex world of how technical progress interacts with public expectations and the realities of the physical world and economic viability. The outcome we take for granted today in the age of mass air transport by jets may look like the inevitable goal toward which the aviation industry was advancing, but the outcome was far from clear in the interwar years. Many dreamed of the eventual goal, but none could be sure of exactly how it would be achieved. It was war which eventually made transoceanic flights the norm in the 1940s – and by littering the world with landing fields made the flying boats redundant. By the 1950s, jets had pushed the speed of aircraft to the point where supersonic flight became a real prospect – although in this case commercial concerns eventually showed it to be impractical for civilian purposes. Jets and rockets had also pushed flight to ever greater altitudes, vindicating some of the early imaginative links between aviation and the next step in the process, space travel. As ever, imagination and practical reality were in constant dialogue.

The Visionaries

There had been much skepticism about the possibility of heavier than air flight in the late nineteenth century, swept away in the excitement generated by the Wright brothers' first flights in 1903. It was the British newspaper magnate Lord Northcliffe who offered the prize for the first crossing of the English Channel won by Louis Blériot in 1909, again resulting in a huge outburst of publicity. By this time, there was intense speculation on the possibilities that aviation would offer for rapid transport around the globe. But heavier than air flight was not the only technology on offer, since the balloons of the previous century were being developed into dirigible airships by pioneers such as Count Zeppelin in Germany. H. G. Wells had speculated about heavier than air machines in some of his early stories, but in his prophetic *War in the Air* of 1908 it was the airship that provided the main means of destruction.

In 'The Argonauts of the Air', Wells had imagined the first flight of a heavier than air machine tested within a huge framework, but finally launched free (due to financial constraints) with disastrous consequences. This is described as the first episode in the struggle for 'man's right of way through the air'. Another story, 'A Dream of Armageddon', depicted war with aircraft, while 'A Story of the Days to Come' had craft with tiers of sails carrying passengers slung beneath in individual seats. The original 1899 version of *The Sleeper Awakes* envisioned 'aeropiles' as means of both transport and war, the name changed to 'monoplanes' in the 1910 reprint. In *The War in the Air*, conflict begins with a German airship attack on the United States and ends with the collapse of civilization as ever-more frightful weapons are employed around the world. Wells imagined more efficient airships using an imaginary technique invented by a single individual, Mr Butteridge. The story also includes small heavier than air 'dragonflyers' and by the time he wrote *The Shape of Things to Come*, Wells had recognized that aeroplanes had become the main form of military flight. In the future war he envisioned, civilization is again brought to its knees by aerial war, but it is the aviators who preserve the spirit of order and progress and begin the reconstruction leading to the first rationally organized society.[2]

Whatever the ingenuity of Mr Butteridge, the airships of *The War in the Air* still got their lift from hydrogen and had a maximum speed of 90 miles per hour. Rudyard Kipling went far beyond this in his 1905 story 'With the Night Mail', in which airships driven by the mysterious 'Fleury's ray' have transformed global communication, flying the Atlantic at a speed of 210 knots. Kipling was honest enough to admit that no one (even in his imaginary future) really understood how the ray worked. Unfortunately for the real-world engineers, it proved impossible to drive airships powered by petrol engines and propellers at any greater speeds than Wells had predicted. In a second story, 'As Easy as A.B.C.', Kipling imagined a world transformed by the availability of personal flying machines. People live isolated lives free to interact only with those they choose to visit, the whole system governed by the rules imposed by the Aerial Board of Control to keep the airways moving.[3]

Between them, Wells and Kipling illustrated the two main reactions to the prospect of aviation: hope and fear. Public response to the news that powered flight had become a reality reflected both of these emotions, the dominant feeling varying from country to country depending on national interests. Kipling focused on the commercial applications and the hope that individuals would soon be able to take to the air in their own machines. These were to become the main enthusiasms displayed in America soon after the Wright brothers' flights. Kipling had spent four years living in New England in the 1890s and the action of 'As Easy as A.B.C.' is set in an America transformed by private aviation. Wells saw the darker prospect of warfare made horrific by the ability to rain destruction onto civilian populations across vast distances. These concerns were more apparent in Europe, traditionally beset with national rivalries, and especially in Britain where the prospect of attack from the air across the barrier of the English Channel was particularly disturbing.

Wells was one of the commentators who saw Blériot's flight as a warning that the Royal Navy could no longer guarantee that Britain would not be attacked. He noted in a London *Daily Mail* article that 'in spite of our fleet, this is no longer, from the military point of view, an island', and the same sentiment was reflected in a supplement to *The Observer*. A few years later, the *Harmsworth Popular Science* serial published reports on developments in aviation, but included an illustration of aerial combat labelled 'The Horror of Warfare in the Air' and another showing a fight between an aeroplane and a submarine. It also described conflicts between aeroplanes and airships – something that would soon come to pass in the Great War when the Germans used their Zeppelin airships to bomb London. Between the wars, the threat of bombing attacks from the air, reflected in *The Shape of Things to Come*, would be a constant fear articulated by novelists and politicians alike.[4]

On the other hand, the British could already see the possibilities of commercial air transport. If aircraft could cross the Channel, the inconvenience of the ferry crossing to Europe could be dispensed with, to say nothing of the expectation for wider links to the Empire and to America. In 1914, the aviation pioneer Claude Graham White teamed up with the aviation correspondent Harry Harper to write a book which ended with the prediction of transatlantic travel with aircraft so fast their wings would have to be reefed. There was a frontispiece depicting the airliner of the future – a triplane with swept-back wings.[5] After the war, designers such as Handley Page used their experience producing bombers to develop the first real airliners, campaigning endlessly to promote the future of commercial aviation. Popular science magazines reported the latest developments and predictions for the future. The daily newspapers routinely reported on long-distance flights and attempts to break speed records. In 1929, the *Daily Mirror* hailed Britain's triumph in the Schneider Cup speed trials, noting that the BBC had broadcast a running commentary. An editorial observed that at the speed achieved, 328 mph, New York could be reached in a day, and speculated that the drive for progress would eventually lead to flights to the moon.[6]

By the 1930s, predictions of an amazing future for aviation became a stock-in-trade of the enthusiast for technical progress. A. M. Low, who had somehow missed the topic in his 1925 book *The Future*, made up for it in 1934 with a chapter predicting mass air travel at speeds of 250 mph thanks to improved designs and wireless-beam guidance. He ended with the speculation that rockets would eventually be used to achieve speeds of up to 3,000 mph at altitudes where meteorites might be a problem. Aviation thus began to merge with the hope of space exploration (see colour plate 5 and Fig. 8.1, p. 133). C. C. Furnas and J. N. Leonard also thought that high altitudes were the key to faster speeds. Furnas noted that the propeller would reach the limit of its performance around 500 mph, but offered no suggestions as to how greater speeds might be achieved. Leonard was concerned that passengers might not want to fly at altitudes where oxygen would have to be supplied. Birkenhead noted that some new means of propulsion would be needed to attain higher speeds and imagined electrically propelled aircraft drawing energy from the ether. Lockhart-Mummery thought that once an aircraft had reached the stratosphere it could hang there and allow the earth to rotate beneath it, presumably a confused image prompted by the idea of artificial satellites. Vernon Sommerfeld also predicted flight in the stratosphere at 500 mph and moved straight on to speculate about space travel.[7]

There was much disagreement over the potential impact of the developments. Many were concerned about the military uses of aviation, but (as with the advent of radio) the optimists predicted that better communication would enhance international understanding and cooperation. Enthusiasts such as Grahame-White and Low took this position. Birkenhead thought these hopes were unlikely to bear much fruit, but suggested that fears about the catastrophic effect of air war were exaggerated. Like many British politicians, he saw air transport as a better way to hold the Empire together in an age of international rivalries. There were suggestions for agreements to place limits on military aviation and even a proposal for an International Air Force.[8]

All of these tensions were visible in the reactions of other European nations. The French were enthusiastic promoters of aviation – huge crowds welcomed Lindbergh when he landed in Paris after his pioneering solo transatlantic flight, even though the country was gripped by a wave of anti-American feeling at the time. The effort to develop long-range airmail routes was a particular source of pride and pilots such as Antoine de Saint-Exupéry gained fame by contributing to the literary celebrations of air power. The Dutch and the Germans also pushed ahead with the development of civilian airlines, although in the latter case the project was increasingly linked to clandestine efforts to restore military aviation following defeat in the Great War. Italy, too, saw aviation as a means of demonstrating its achievements on the international stage, using flying boats to show the potential for long-distance routes around the world. For the Fascists, however, aviation had a more sinister political message. As Benito Mussolini wrote in 1923: 'Not everyone can fly . . . Flying must remain the privilege of the aristocracy; but everyone must want to fly, everyone must regard flying with longing. All good citizens, all devoted citizens, must follow with profound feeling the development of Italian wings.'[9]

In America, it was the sense of personal freedom implied by the opportunities of aviation that drove the early waves of enthusiasm. Here, there was initially no threat of attack from abroad and a vast area that could be accessed by improved transportation. In 1919, the *New York Times* saw an aviation exhibition revealing 'the threshold of a new age whose developments the most imaginative can hardly imagine'. An almost religious atmosphere surrounded the movement – aviation had a miraculous potential to transform people by empowering them through the freedom to travel anywhere they desired. The prospect of personal aircraft was especially attractive. Heroes such as Lindbergh hailed their achievements as pointers to developments that would open up the world for all, and like many other enthusiasts Lindbergh linked his passion with the future world of rockets and space travel. In 1934, a popular aviation magazine introduced a regular feature on 'Planes of the Future' and used some of the imaginary designs as cover art.

For these optimists, the pioneering flights, whatever their dangers, pointed the way toward what would become the mass transportation of the future. They were also convinced that the resulting cultural interactions would help to generate world peace. Harry Guggenheim, a friend of Lindbergh who used the family fortune to bankroll a Fund for Promotion of Aeronautics, wrote that aviation would lessen the distances between cultures so that '[o]ne by one the barriers between nations are falling'. Unfortunately, in view of later events, one of his prime examples was the modernization of Japanese culture. Richard Byrd thought the pioneering flights 'assist the great progress of the race towards its goal'. Aviation was an instrument of peace as well as progress, and he predicted that transoceanic flights 'will bring us closer to the nations of the world both in distance and in sympathy'.[10]

Girdling the Earth

The pro-aviation lobby had huge ambitions, but no clear idea how best to achieve them. In fact, technical developments were slow in the 1920s and air-travel did not catch on with the public. One crucial issue was the lack of consensus on the best technology to use. A 1927 article in a British aviation magazine repeated the litany that record-breaking flights pointed the way to the standard performance of tomorrow and predicted transoceanic flights. But it conceded that it was not clear whether these would be achieved by aeroplanes (possibly using mid-ocean 'floating islands' to refuel) or flying boats. Since the magazine was *The Aeroplane*, the article didn't even mention airships, which had also already flown the Atlantic. Two years later, Sir Charles Dennistoun Burney provided a detailed survey of the prospects for both heavier-than-air machines and airships. Like several other commentators, he suspected that some combination might be necessary to make the overall system work.[11]

Through into the late 1930s, the proponents of three rival technologies strove to improve their systems and convince governments and the public that theirs was the best way forward. The debate over how to establish regular transoceanic air transport would only be resolved when World War II led to massive improvements in land based aeroplanes as bombers. The war also introduced the jet engine, which fulfilled the prediction of much faster speeds.

The three technologies all faced similar problems, but each had advantages in some areas and disadvantages in others. The one common advantage was that aircraft could cross barriers that were obstacles to land-based transportation. Small bodies of water were no problem, as Blériot demonstrated for the English Channel, and undeveloped territory lacking roads or railways was also a prime target for air transport. Not surprisingly, one of the first success stories in civil aviation was the route between London and Paris, cutting out the transfer from train to ferry and back again. The situation was very different over areas already supplied with continuous rail links. The early aircraft were not much faster than express trains and could not depart from city centres (although efforts were made to overcome the latter problem). More speed was needed to convince the public of the advantages, but there were also the questions of safety and comfort. Night flying was dangerous, and aircraft were much more likely than trains to be delayed by bad weather. The early aeroplanes were noisy, cramped and offered no heating. They also had very limited range, making transoceanic flights impossible. Airships might be slow and cumbersome, but they were at least large enough to carry passengers in comfort and held enough fuel for long-distance journeys.

To the enthusiasts who read *The Aeroplane*, it was obvious that heavier-than-air machines were the way forward, at least over shorter distances. As the British designer Handley Page argued in 1919, airships were too big and too expensive both to build and maintain. He had built bombers during the Great War and was now using these designs as the basis for new two- and four-engined civilian machines for the emerging airline companies. But a year later, businessmen were still reluctant to switch to air travel even over the London–Paris route. In 1925, the magazine conceded that across Europe, where rail travel was fast and convenient, take-up of air travel was limited. Nevertheless, there was some progress: regular services between London and several European cities had been established and in 1924 Imperial Airways began services to India (it took twelve days to get there). The government was convinced that air transport would revitalize the Empire, a sentiment echoed by Burney: 'the advent of the air age will see the inauguration of a bold and comprehensive policy of Empire development worthy of the glorious traditions of our race'.[12]

All across Europe, airlines were being created, the French, Dutch and Germans being especially active because here governments were more prepared to lend financial support. Airlines whose names are still familiar today stem from this period, including KLM and Lufthansa (see colour plate 7). Carrying mail was an important source of revenue for these early projects at a time when passenger numbers were limited. This was also true in the United States, where developments were limited in the early 1920s. By 1930, things were improving: several airlines were now offering transcontinental flights (via intermediate stops) and businessmen were starting to see air travel as normal.[13] Here, too, airlines were founded which dominated the industry for decades to come: United, Eastern, TWA and the like.

What made all this possible was a series of technical developments increasing the speed, size, range and reliability of aircraft. Better engines were developed, and as speeds increased, the ancient biplanes were replaced by monoplanes built of aluminium alloys using stressed-skin construction which eliminated the need for struts and bracing wires. Aerodynamics and the use of wind tunnels showed the advantages of streamlining at the speeds being achieved, so the ungainly ex-bombers were replaced by the smooth lines of aircraft such as the Douglas DC 3, introduced in 1935. Imaginative designers such as Norman Bel Geddes published images of huge futuristic aircraft with vastly increased speeds. Speeds were in fact beginning to increase, although the dreams of the visionaries were still out of range. One expert was surprisingly pessimistic – in 1929, Nevil Shute thought that it would take fifty years to develop an aircraft that could carry a five-ton payload at 95 mph for 200 miles. Only two years later, William Forbes-Sempill urged businessmen to make use of the opportunities offered by aviation, noting that at 95 mph the speed of a commercial aeroplane was twice that of the average express train. By the late 1930s, this had been doubled by planes such as the DC 3, which could carry twenty-one passengers across the United States from coast to coast in less than a day.[14]

Navigation could be difficult, especially in poor visibility. There was no thought of anything like radar, although Low and others foresaw the use of radio beacons that would allow pilots to locate themselves. Guggenheim predicted a radio direction-finder and noted the value of radio to alert pilots to poor weather ahead, lamenting that America lagged behind Europe in the provision of weather information. He urged every town in America to paint its name in large letters on a prominent building such as the railroad station. A system for radio guidance to permit landing in fog was proposed in 1938 and during World War II the Germans guided bombers to their targets by radio beams. Night flying was also difficult. Runways could be illuminated and in densely populated areas such as northern Europe there were plans for networks of searchlight beacons at airports and city centres (Figure 7.1). By the late 1930s, crossing continents (at least those densely populated) had become unproblematic as long as the weather held good.[15]

One drawback for the traveller in a hurry was that airfields had to be situated outside city centres, both for safety reasons and because they occupied a considerable area. The problem was compounded by the necessity to take

Fig. 7.1. Proposed system of searchlight beacons to allow night flying between Paris and London, from A. Christofleau, *Les dernières nouveautés de la science et de l'industrie* (1925) p. 5.

off and land into the wind to reduce the risks created by excessive ground speeds. To deal with this limitation, there were numerous proposals for schemes that would create runways above the city, on the roofs of skyscrapers or perhaps on bridges linking them far above the streets below (see colour plate 6). Perhaps the most ambitious was Charles W. Glover's plan to build an aerodrome above the King's Cross railway yards in central London, a model of which was exhibited at the Institute of Civil Engineers in June 1931. This would have a circular perimeter and either four or eight crossed runways so that planes could always land head-to-wind. Passengers would take lifts down to the ground where local or national rail transport was to hand. Another possibility was a rotating airstrip mounted above a skyscraper so that it could be turned into the wind. Geddes proposed a rotating airport to be built in New York harbour. But as airspeeds increased, so did landing speeds, making such schemes always one step behind technical feasibility. As late as 1946, designs for new airports outside Paris and New York included multiple runways radiating out from the central terminal to eliminate the problem of cross-winds.[16]

In his contribution to the 'Today and Tomorrow' series, Oliver Stewart complained of the 'tiresome persistence' of those who urged the totally impractical idea of rooftop airstrips. He insisted that the only way of bringing aviation into the city centre was what would later be called a vertical take-off and landing system (VTOL). He joined a chorus of enthusiasts calling for the development of such a system as a matter of urgency. One obvious solution was the helicopter, which had been investigated by inventors including Thomas Edison since the late nineteenth century. Experimental designs were being tested in the 1920s and optimists such as Low confidently predicted their introduction. Lockhart-Mummery thought the London of 2456 would have helicopter taxis. But the technical difficulties were enormous and the first viable machine did not appear until 1936. This was in Germany, although development then moved to the United States with the work of Igor Sikorsky. By 1944, *Popular Mechanics* could more realistically label the helicopter as the 'air bus of the future'.[17]

In the meantime, Stewart was one of many who saw the autogyro as the only practical way to achieve VTOL capacity. As developed by Juan de Cierva in 1923, the autogyro simply replaced the fixed wing with large freely rotating blades. It did not give true VTOL, but could take off from and land on a very restricted area. The invention was taken up in both Britain and the United States and actually came into limited use for military purposes and mail delivery. Commentators who doubted the viability of the helicopter accepted the autogyro as a practical alternative and many aviation enthusiasts saw it as the way forward that would allow air travel to work directly from city centres.

As late as the 1950s, British engineers tried to marry the principles of the helicopter and the autogyro in the Fairy Rotodyne. This used tip-jets to power the main rotor for taking off and landing vertically, but during horizontal flight with turboprops these were switched off, converting to autogyro mode. The craft was tested in 1957 and demonstrated at the Farnborough air display, hailed as the future airbus that would operate directly from city centres yet have speeds of up to 150 mph. Unfortunately, the tip-jets were so noisy that no one wanted them operating in their neighbourhood and the project was soon cancelled.

Flying Boats or Airships?

Between the wars, many thought that aviation would develop on a two-stage basis: autogyros would ferry passengers from the city to the nearest airport, where they would transfer to long-distance flights. The crucial question was then to decide which technology offered the best prospect for the long-haul part of the journey.

Aeroplanes could cope with long distances only by making frequent refuelling stops, which was feasible for overland routes, but left the oceans as major barriers. They might not be much faster than express trains, but they were certainly faster than ocean liners. Yet land-based aeroplanes could only make transoceanic flights at the limits of their endurance. Alcock and Brown flew the Atlantic in 1919 and Lindbergh made the first solo crossing in 1927, but they had flown with no margin for error as far as fuel was concerned and there was certainly no room for passengers or even mail. Years of technical innovation would be required to overcome this limitation and in the meantime there were two more promising alternatives. Flying boats could be built much larger than land-based aeroplanes and hence could carry more fuel (they also had obvious safety benefits on ocean crossings). Airships could also be constructed big enough to carry the fuel needed for long voyages, and could even carry passengers in comfort. They were not as fast as fixed-wing planes, but they could still make an ocean crossing in half the time of a liner.

The desperation of the aircraft designers is evident from the plans that were circulated to moor way-stations in mid ocean to allow for refuelling and maintenance. In 1927, the Germans used a ship in the south Atlantic to refuel seaplanes making the crossing from Africa to South America, upgrading this later to a larger vessel capable of winching the planes on board. The boldest proposal came from American engineer Edward R. Armstrong, who had built a giant platform to carry machinery for extracting bromine from sea water. In 1929, he proposed using the same technique to construct floating runways in the north Atlantic on which land-based planes could land. These would be 900 feet long and would carry repair works, a meteorological station and even a hotel where passengers could take a break. They would stand above giant flotation chambers that would ride beneath the surface where there is no wave action. Platforms would be moored every 500 miles across the main ocean routes. The proposal attracted considerable attention in Britain, while the Germans made a movie, *F.P.1 Antwort Nicht* in 1933, in which the action was set on such a Flugzeug Platform (a mock-up was built in the Baltic). An English-language version was also produced under the title *F.P.1 Doesn't Answer*.[18]

Armstrong's platforms were never built because two alternatives emerged that could provide non-stop transatlantic services. The ships stationed to allow the refuelling of seaplanes were a temporary measure soon made unnecessary by the construction of huge flying boats with a much greater range. At this time, flying boats were as fast as land-based aircraft and had the advantage that they could land on any convenient stretch of water. European nations wanted aviation to link their far-flung empires, but constructing airfields in remote and undeveloped regions was expensive and flying boats offered a cheaper alternative. Because they didn't need landing gear, there was a considerable saving in weight. Hulls could be larger because their weight was spread more evenly when afloat. Giant flying boats could thus have significantly greater range than land-based aircraft and could even provide a modicum of comfort for passengers. The British decided that flying boats would be a viable alternative to Imperial Airway's land-based craft and promoted the construction of the Shorts Empire machines, which served routes to India, Australia and South Africa through the 1930s. Considerable romance surrounded these flights, although there were numerous accidents.

The early models of the Shorts boats still could not cross the Atlantic, so in 1938 a scheme was developed by Major Robert H. Mayo of Imperial Airways to piggyback a seaplane on top of a flying boat. A modified Empire boat, *Maia*, carried the seaplane *Mercury* up to a significant altitude and some way out to sea, at which point *Mercury* would detach and complete the journey. The system worked, and even broke the long-distance record with a flight to Cape Town, but it could never have been used for carrying anything but mail. The French laughed at the combination, partly because it was so obviously a stopgap, but also because the Americans were already building flying boats that could cross the Atlantic in one trip. By the mid-1930s, Sikorsky's S40 Southern Clipper was flying forty passengers to the Caribbean and Central America and soon the China Clippers were flying to Hong Kong. In 1939, the Boeing B-134 began services across the Atlantic, although it had to make intermediate stops in Newfoundland, the Azores or Ireland. A later version of the Shorts boat

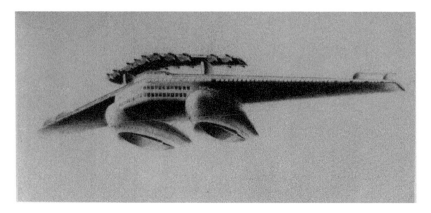

Fig. 7.2. Norman Bel Geddes's design for a giant flying boat. From D. Bush, *The Streamlined Decade* (1970), p. 27.

was introduced to compete, but flights were soon suspended due to the outbreak of war in 1939.

The Clippers became part of the mystique of the streamlined age, and popular science publications were filled with images of even bigger and more futuristically styled flying boats. Norman Bel Geddes proposed a twin-hulled machine that could carry huge numbers of passengers on intercontinental routes (Figure 7.2). Ritchie Calder's *Birth of the Future* had a similar design by a German company for its frontispiece and as late as 1942 *Practical Mechanics* imagined freight transport revolutionized by containers loaded into flying boats. For a brief period, the flying boat thus came to represent the best way forward for long-distance heavier-than-air flight. The vision did not survive the next war: by 1945, heavy bombers could routinely fly the routes that had challenged the flying boats and the world was covered with airstrips. The British alone tried to continue the technology into the 1950s. The Saunders-Roe Princess flying boat had ten turboprop engines and was supposed to be adopted by British Airways. The author remembers as a boy seeing the forlorn structure of the prototype mothballed on the Isle of Wight after its test flights.[19]

There was another alternative that could easily fly the Atlantic non-stop by the 1930s: the airship. Some supporters of heavier-than-air craft conceded that airships offered the best chance of conquering the transoceanic routes, while others were bitterly critical. E. F. Spanner wrote a series of books attacking the airship industry, one engagingly titled *Gentlemen prefer Aeroplanes!* Spanner also produced a two-volume analysis of the crash of the R101 which ended the British airship programme. Yet, writing three years after that event, Captain J. A. Sinclair insisted: 'The truth of the matter is that aeroplanes have no chance whatever against airships in the service of long-distance overseas routes. There can be no question at all of the fact that the future of world trade is bound up with the future of airship services.'[20] Sinclair attributed the catastrophe of the R101 to the fact that it had been rushed into service for political reasons and complained that the Germans were now forging ahead with their Zeppelin machines. Despite a series of setbacks, the airship industry flourished for a brief time, during which it did indeed create the first commercial transatlantic services. The United States saw the airship more as a naval scout, producing two ships, the *Akron* and the *Macon*, which could launch and retrieve light aircraft to widen the range of ocean surveyed.

The main advantage of airships was that they could be built to a gigantic size, giving them a much greater carrying capacity than any heavier-than-air machine of the period. They could hold enough fuel for long-distance flights and still carry their passengers in relative luxury. The latter factor was important because among their many limitations was the fact that they were relatively slow, with a cruising speed usually of around 70 mph. Against a strong headwind, their speed over the ground was little more than a crawl, and the main skill of an airship captain was finding the altitude with the most favourable wind speed and direction for the voyage. Low noted that they were unlikely to get any faster, which meant that in the end aeroplanes were almost certain to win the race. Airships' sheer size meant that they were slow, fragile and cumbersome, easily damaged by rough weather and they needed large crews on the ground to manoeuvre them when mooring. They also needed vast sheds for maintenance, although masts were developed for short-term mooring (and there were hopes that these could be installed on the tops of skyscrapers). Finally, there was the problem of the gas used to provide lift, usually hydrogen, which was highly flammable. Helium could provide an inert alternative, but it was a scarce by-product of the American oil industry, and the Americans were reluctant to sell it abroad.

Fiction writers speculated about new gases that might solve the problem, but these were as imaginary as Kipling's mysterious 'Fleury's ray'. Some went further and imagined new technologies that would simply annul the force of gravity. Wells had imagined a substance that could do this in his *The First Men in the Moon* and the idea became a staple of science fiction between the wars. But it also found its way into popular science periodicals as though it were a real possibility. In 1929, *Popular Mechanics* depicted an airship of the future that floated with anti-gravity technology. In 1942, the British *Practical Mechanics* published an apparently serious article claiming that the basic principle for such a system had already been worked out and picturing a heavily armoured military aircraft with no wings. John W. Campbell later argued that studying atomic physics might eventually reveal how gravity might be controlled, marking the end of aeroplanes.[21]

In the meantime, airship designers had to get their lift from gas, and for most this meant hydrogen. The many potential dangers were ignored by a vocal lobby proclaiming that airships represented the future of long-distance aviation. As early as 1909, the *Illustrated London News* published an image of a proposed giant airship for the Royal Mail. In 1919, the British R34 crossed the Atlantic only a few weeks behind Alcock and Brown. The British programme was halted, temporarily as it turned out, by the crash of the R38 in 1921. Nevertheless, popular science publications enthused over the prospects for regular transatlantic services and in 1929 Burney lent his support, arguing for the development of an airship capable of landing on water to eliminate the difficulties of mooring on land.[22]

By this time, the programme had been revived thanks to the efforts of the politician Christopher Thomson, later Lord Thomson of Cardington (where a government factory had been established). He envisioned airships as an alternative to Imperial Airways' aeroplanes for transport around the Empire. Two ships were built, the R 101 at Cardington and the privately financed Vickers R 100. Vickers's chief designer was Barnes Wallis, later famous for inventing the bouncing bomb used in the dambusters raid. His team included Nevil Shute Norway, who would achieve fame as a novelist under the pen-name Nevil Shute. The R 100 was the more conventional machine and flew the Atlantic successfully to Montreal and back. The R 101 was more experimental and underwent a series of modifications in the course of development.

In October 1930, Thomson insisted that the R 101 be made ready to take him to an Imperial conference in India. The ship left in bad weather without being properly tested and crashed in the night on a hillside in northern France, killing Thomson and almost everyone else on board. The *Daily Express* report of the tragedy argued that the claims of the airship lobby had been swept away, vindicating the experts who favoured heavier-than-air machines. The British programme was cancelled and the R 100 scrapped. Shute's autobiography is bitterly critical of the failings of the government-funded group, although there remained some enthusiasts who insisted that if it had been allowed to develop properly, the R 101 would have revolutionized the industry.[23]

The Americans also lost airships: the *Shenandoah* in 1925, the *Akron* in 1933 and the *Macon* in 1935. This left the field largely to the Germans, whose Zeppelin machines were flying with apparent success. The Graf Zeppelin flew a number of long-distance voyages, including regular flights to South America. In March 1936, the *Hindenburg* was launched with a huge publicity campaign (see colour plate 8). It provided luxurious accommodation and began a regular service to the United States. The Americans were especially reluctant to supply helium to the Nazi regime, which was using the airship for propaganda, so the *Hindenburg* still got its lift from hydrogen. On 6 May 1936, it burst into flames at Lakehurst Naval Air Station, a catastrophe recorded by an army of photographers and a live radio commentary.

Despite the negative publicity, the Germans vowed to continue their programme and there were calls for an *entente* between America and Europe that would allow wider use of helium. But for all practical purposes, the Hindenburg tragedy brought the era of the airships to a close even before the outbreak of the war that would stimulate the development of aeroplanes with a size and range that the previous generation could only have dreamt of. Whether the airship could still have played a role in a world that did not experience the effects of the war is an interesting possibility for students of counterfactual history to ponder. The episode certainly drives home the fact that in the mid-1930s it would have been hard to predict which technology would succeed.[24]

The Anti-Aviation Lobby

The disasters experienced from time to time by all forms of aviation hardly encouraged the public to take up the new form of transport. Only those anxious to save time would be enthusiastic. Nor did the disputes between the proponents of rival technologies inspire confidence. Some commentators were even more skeptical about the industry's potential for improving the world. There was the ever-present worry that aircraft would make war even more horrific. Some called for international control of aviation to prevent its misuse. A few even advocated the complete destruction of all aircraft on the grounds that even civilian machines could be adapted for war. As with radio, claims that better communications would promote world peace were greeted with derision by many, including George Orwell. Wild predictions of speeds up to 1,000 mph were dismissed as impractical, if only because the human body could not stand the stresses created by manoeuvring at such speeds. As late as 1937, Sir Harold Harley told a BBC audience that no major innovations could be foreseen in aviation technology.[25]

Amid this chorus of suspicion, there were a few voices arguing that the whole enterprise was driven by wild exaggerations of the potential usefulness of air transport. These critics seem to have been most active in Britain. They suggested that the aviation industry was a scam perpetrated on a gullible government by self-styled experts desperate to preserve their jobs or maximize their profits. Aviation could never operate effectively unless it was supported by endless subsidies, and even then its future development would remain hamstrung by problems that were brushed under the carpet to create an illusion of progress.

The most persistent critic was an ex-Royal Navy submariner, Commander Bernard Ackworth. We have already encountered Ackworth's resentment of new technologies in his call for the Navy to abandon oil as a fuel because Britain had to depend on overseas sources. His opposition to aviation was in part driven by a desire to preserve the Navy as Britain's key line of defence, a position challenged by those who emphasized the threat of attack from the air. But he was a conservative thinker on a wide range of positions, an evangelical Christian who became a founder member of the Evolution Protest Movement. In his *This Bondage* of 1929, he managed to link his opposition to aviation and Darwinism via a curious argument based on the difficulty of allowing for cross-winds when navigating in the air. Birds were unable to do this properly and had to make constant course corrections when flying toward a perch. This meant that animals were devoid of intelligence, so the human mind could never have evolved from an animal ancestry. But the same problem also applied to all forms of aviation. The pilot of an aircraft could never be sure how strong any cross-wind might be, and it was thus impossible to plot a course directly toward its destination.

Ackworth had published this argument in *The Spectator* two years earlier and his article was noted in the most comprehensive attack on the aviation industry, *The Great Delusion*, published under the pen-name 'Neon' in 1927. It turned out that 'Neon' was actually Marion W. Ackworth, the wife of Bernard's brother (who had founded the Ilford photographic company). Whether Marion actually wrote the book or merely allowed her name to be used to conceal Bernard's naval background is a matter of conjecture. The book articulated a barrage of complaints about the failure of the aviation industry to substantiate its claims both for its military applications and for the prospects of a financially viable commercial airline system. The government was pouring money into the industry, deluded by the promises of those who sought only to profit from it. *The Great Delusion* received some publicity in the British press and elicited a book-length rebuttal by the aviation enthusiast Rear-Admiral Murray F. Sueter. Arthur Hungerford Pollen, who had written a preface endorsing Neon's arguments, repeated similar objections in the science magazine *Discovery*.[26]

Needless to say, the campaign did little to block the progress of aviation technology. Ackworth's argument that it was impossible to allow for cross-winds when navigating an aircraft sounds ludicrous today, but it seemed more plausible at a time when airspeeds were less than 100 mph, and it certainly had some validity for flights in poor visibility when it was impossible to check movement over the ground. In fact, the aviation industry was already predicting the development of aids such as radio beacons that would provide secure guidance, although no one anticipated the development of radar. Ackworth could not conceive of technical solutions to his problem in part because he was suspicious of the whole ideology of scientific progress. In a later book, *The Navies of Today and Tomorrow*, he attacked the scientists and technical experts who were systematically misleading the government to sustain their industries and careers at public expense. Science was merely factual knowledge, but the emergence of a professional research community had allowed cranks and enthusiasts to promote wild theories linked to expensive and often unworkable technological innovations. 'Freakishness, excess and abortion are the hall-marks of this queer Progress', he insisted, and it was time to take stock of the research craze that had begun to distort the professions of engineering, medicine and the like.[27] Some literary figures may have agreed with this assessment, but industrialists and politicians remained unconvinced.

Personal Flying

At the opposite end of the spectrum of opinion were the enthusiasts who expected that soon everyone would be able to fly their own personal aircraft. Very light aeroplanes were already being used by amateur pilots in the 1920s, but these still needed a small field from which to operate. Henry Ford sought to tap into this potential market with his Flivver, introduced with a blaze of publicity in 1926 as 'the Model T of the air'. Charles Lindbergh was brought in to test the machine, later confessing that it was the worst plane he'd ever tried to fly. In 1928, a flivver being tested by Ford's friend Harry J. Brooks crashed, killing him and leading to the cancellation of the whole project. The idea must have appealed to Aldous Huxley's imagination, though, appearing in one of the mantras recited in his *Brave New World*: 'Ford's in his flivver. All's well with the world.' Despite Ford's setback, the US Government spent half a million dollars in the early 1930s trying to develop a cheap, easy-to-fly light aircraft. Various projects were tested, including some that involved cars that could be converted to planes by adding on wings. In the end, however, light aircraft remained useful only to those who had the space for landing and take-off.[28]

The enthusiasts who believed that everyday life would be revolutionized by aviation realized that some form of VTOL system would be needed if the ordinary citizen was to take off from his or her own garden or apartment roof. As early as 1928, *Popular Mechanics* predicted a car that could be turned into a helicopter, but most commentators thought the autogyro was a better bet – although it did need a short horizontal run before take-off. An article in *Armchair Science* in 1932 noted that there was still considerable public skepticism about such machines and they were never developed seriously for the individual flyer. Nevertheless, the technophiles continued to hope that some form of VTOL technology would solve the problem and get the public into the air. In 1947, designer Walter Dorwin Teague predicted that cars would be replaced by light aircraft. In 1951, *Popular Mechanics* pictured a mini-helicopter being pushed out of a suburban garage, while in 1957, it predicted a flying fan vehicle that would become available in ten years' time. As late as 1971, Isaac Asimov was still expecting that VTOL machines would eventually take the place of automobiles.[29]

An even bolder proposal was to adapt the rocket, now being developed by a number of inventors, to provide a means of personal flight. Only the very brave would relish the prospect of blasting off with a rocket strapped on their back, but the prospect of what became popularly known as the jetpack seems to have caught the imagination of many young men at the time. It was suggested in various science fiction stories and became the basis for a twelve-part movie serial *King of the Rocket Men* in 1949 (the author remembers seeing this as a boy some years later). The failure of the jetpack to materialize in the 1960s was a source of much frustration for a generation of youngsters obsessed with technology as a source of personal liberation.[30]

The failure of the aviation industry to satisfy the demand for personal aircraft is used by Garry Westfahl to illustrate the fifth of his 'fallacies of prophecy': the assumption that a new technology will follow the same pattern of development as the previous one in the same area.[31] People took it for granted that aviation would follow motoring in spreading out to become available to all. It didn't happen in part because flying is more complicated than driving an automobile and got more complicated as machines became more sophisticated. Most people don't have the time or the dedication needed to fly a plane safely (let alone a helicopter), even if they could afford one. The very rich now use helicopters for personal transportation, but most of them hire professional pilots. There is also the problem of what we now call air traffic control. The carnage that would have resulted if thousands of people had taken to the air over our cities would have dwarfed that which occurred in the early days of motoring. As Asimov conceded in his 1971 prophecy, personal planes would have to be guided along predefined paths by radio beacons, and in the end the whole system would have to be automated. This would, of course, eliminate the sense of freedom which drove the expectation in the first place.

The Jet Age

Individual citizens may not have acquired their own planes, but the years following World War II saw the fulfilment of at least some of the earlier prophecies. Under the pressure of military necessity, long-range aircraft were developed which could safely fly across the Atlantic. The first airliners of the post-war era were often just modifications of these long-range bombers. Speeds were increasing, too, thanks initially to the introduction of the turbo-prop. British designers were convinced that they could beat the Americans in the production of a new generation of airliners. The Bristol Britannia was conceived in the late 1940s to carry 100 passengers across the Atlantic at speeds up to 400 mph. Bedevilled by the technical and political problems that were all too typical of the British aviation industry after the war, it was not introduced until 1956. By this time, the first generation of jet airliners were starting to appear, carrying more passengers at speeds of over 500 mph. The De Havilland Comet entered service in 1952, although a series of disasters due to structural failures forced it to be withdrawn until a modified version re-appeared in 1958. This was too late: American jets such as the Boeing 707 had already begun to corner the market, carrying more passengers more economically. By the time Asimov was still trying to revive the hope of personal planes in 1971, an ever-increasing number of passengers were being transported around the world in airliners at speeds the previous generation had only dreamt of.[32]

The jet engine was a new technology that had not been foreseen by the prophets of the pre-war years – they had pinned their hopes mostly on rockets. Developed by both British and German engineers, the jet came into service in the last stages of the war, powering the Gloster Meteor and the Me 262. The Meteor went on to establish new speed records in the post-war years, by which time British, American and Russian teams were racing to develop military jets to fly at ever greater speeds. The move to jet-powered civilian craft was an obvious extension of what had begun as a purely military programme.

The civilian jets are turbofans designed to operate at speeds around 500 mph. Military specifications soon led designers to aim for aircraft that would operate beyond the speed of sound, pushing toward the predictions of 1,000 mph suggested by some pre-war optimists. This involved breaking what came to be known as the sound barrier, for some time conceived as a limit beyond which planes could not pass without disintegrating. Propeller-driven planes could never hope to reach the speed of sound, but wartime pilots approaching it in dives had experienced violent buffeting and unpredictable behaviour. German technicians had studied this problem in the 1930s and had realized that new wing configurations would be needed to cope with supersonic speeds, most obviously the swept-back wing. They produced a plethora of futuristic designs, most of which were never built.

Post-war military jets soon began to approach the speed of sound, their experiences generating the myth of the sound barrier. British designer Geoffrey de Havilland's son was killed in 1946 testing one of the firm's prototypes produced to test the new wing configurations at high speeds. David Lean's 1952 film *The Sound Barrier* was loosely based on this tragedy and helped to cement the myth in the public imagination. Walter Kaempffert was one of many popular science writers who described the problem for American readers. Yet, by then, the barrier was already being broken regularly by military jets (although the first plane officially to do this was the rocket-powered Bell X-1). Supersonic flight was certainly noisy, but it was no longer seen as inherently dangerous.[33]

The visionaries of the pre-war years assumed that progress toward these immense speeds would eventually be applied to civilian travel. Kaempffert had reinforced these expectations in 1940, predicting that within fifty years one would be able to have breakfast in New York and lunch in London. In fact, it was in the 1960s that plans for a civilian supersonic transport began to take shape in both America and Europe. The *Boy's Own Paper* told its readers of a design by Barnes Wallace for a swing-winged passenger jet that could fly to Australia at a speed of 1,700 mph. It conceded that there was no financial backing for the plan and warned that the Americans were interested in the idea. In 1964, *New Scientist*'s predictions for twenty years ahead had included the claim that by then most long-range flights would be supersonic.[34]

This prediction was by a member of an Anglo-French team now testing a new wing design for what eventually became the Aérospatiale-BAC Concorde. Two years earlier, though, *New Scientist* had noted that enthusiasm for the project was confined to the engineering companies and to governments – neither the airlines nor the public saw the supersonic option as a priority.[35] Driven by the desire for national prestige, Concorde first trialled in 1969 and entered service in 1976, encountering considerable opposition in the United States, where the parallel Boeing 2707 project had been cancelled in 1971 in part because of the noise problem. Concorde survived until 2003, when it was withdrawn for economic reasons. Supersonic flight was attainable, but it was so expensive that only the very rich could afford it and when the economy was weak there simply weren't enough passengers to make the service viable. Here, the optimists' predictions were confounded not by technical limitations, but by social and economic factors.[36]

Concorde did, however, set aside one important aspect of post-war technical enthusiasm: the vision of superfast planes driven by atomic power. From the earliest days in which the public was made aware that atomic energy was at last becoming a reality, the expectation was encouraged that it would offer immense benefits for everyday life. Among these visions was that of an aviation industry based on atomic powered jets or rockets. Through the 1940s and 1950s, popular science books and magazines abounded with predictions that atomic power would allow the construction of aircraft of amazing speed and endurance. In 1953, *Popular Mechanics* reported test-pilot Roland Falk's prediction that within fifty years there would be two-hour flights between London and New York, the planes never actually landing to refuel so that passengers had to be ferried up to them by air taxis. The British boys' comic *Eagle* depicted an airfield of the future with atomic powered rocket planes capable of flying at over 1,000 mph. It never happened, of course. Already by 1947, John W. Campbell was one of a growing number of skeptics pointing out that the weight of shielding required would make atomic-powered planes impractical (although he conceded that remote-controlled ones might be operated). The United States did create an Aircraft Nuclear Propulsion Program in 1946, but it was eventually wound up in 1961.[37]

There have been a few even more imaginative projections. One idea centres on the emergence of new physical principles that would allow the effect of gravity itself to be counteracted. Several novels in the interwar years imagined the development of anti-gravity devices. Shaw Desmond's *Ragnarok* had 'magneto-wireless engines' that tapped the Earth's magnetic field to allow a craft to hover and in Michael Arlen's *Man's Mortality* the invention of 'Feaveryear's anti-gravitation gyroscopic stabilizer' makes all conventional aircraft obsolete. Projects aimed at achieving anti-gravity were actually established by several organizations in the late twentieth century, including Boeing, NASA and Britain's BAE Systems. Nothing has come of these ideas, although the hope of countering gravity has remained a staple of science fiction since Wells imagined it for his *First Men in the Moon*. In 1962, Arthur C. Clarke predicted both gravity control and instantaneous transportation by scanning material objects so the information defining their structure could be transmitted and the structure reconstructed elsewhere. This would become the 'transporter' of the *Star Trek* television series, although Clarke realized that it would actually allow for duplication of the scanned object.[38]

Leaving such extravagances aside, the successes and failures of the predictions about aviation made in the earlier decades still illustrate the uncertainties involved. The visionaries were inspired by the optimism of the engineers and technicians and they were certainly pointing roughly in what turned out to be the right direction. But getting the details right was problematic, as the conflicts over the best way forward in the 1930s revealed. The expectation of mass air transport at high speeds was eventually realized, but was made possible in part by unforeseen inventions conceived and developed by the military. The jet engine and radar were to revolutionize air transport, but only after they had made their mark in war. Some of the bolder projections have never been actualized, not because the technologies cannot be developed, but because they are not commercially viable. We do have helicopters, but not for every man, and the only supersonic aircraft are used for military purposes where economic factors do not count.

8   Journey into Space

The most adventurous prophets of technological progress assumed that the development of aviation would lead eventually to the exploration of the moon and the rest of the solar system. Aircraft, probably powered by rockets, would fly ever faster and higher. It would be a short step from the transatlantic rocket to a space station and then to the moon. In his 1928 contribution to the 'Today and Tomorrow' series on exploration, J. Leslie Mitchell praised the aircraft's ability to survey the earth, but also included a chapter on space travel. He insisted that: 'Given ten years, a well-equipped laboratory and a competent staff of assistants, there is hardly an expert artillery officer, well-grounded in chemistry, but could succeed in achieving inter-planetary communication with experimental projectiles.' In the following year, the *Daily Mirror* hailed Britain's success in the Schneider Cup air race by declaring that: 'Having got into the air man will not rest content with one planet . . . It may not be very long before humanity will realize the dream of reaching to the moon.'[1]

Fired by similar enthusiasm, a dedicated band of enthusiasts strove to create a viable rocket technology. Their work was greeted with derision by more conservative thinkers (even within the sciences) and their success was limited. No commercial enterprise would take the project seriously and (as with aviation itself) it was only the pressure of war that led to rockets becoming a workable technology. In the post-war decades, rockets became symbols both of terror and enterprise, the fear of the nuclear missile being balanced by the first real steps into space and to the moon. The 1950s and 1960s saw a more widely based flurry of enthusiasm, sadly let down when it turned out that the Apollo missions were only a reconnaissance driven by Cold War tensions.

Whatever the limitations of the rocket programmes, they did help to generate enthusiasm among a small but dedicated band of technophiles. Here more than in any other field science prophecy and science fiction blended in a complex interaction. The enthusiasts who read popular science magazines also read science fiction, and some well-known authors wrote in both genres. Science fiction, as it spread beyond the pulp magazines to radio, cinema and eventually television, alerted the public to the prospect of space exploration. But relations were not always smooth. The rocket-builders were happy enough to see fairly plausible representations of their plans in what is now known as 'hard' science fiction. Here the story is set in a future society exploiting technologies that seem feasible and in some cases are already being considered. But the more exotic space operas offered to the public seemed little more than caricatures in which badly conceived rockets transported all too conventional heroes and villains. Even the most respected science fiction authors moved on to create futures on a vast scale that allowed them to muse on the moral and spiritual possibilities open to humankind. The link between this kind of science fiction and the space programme was problematic. Ordinary readers might see it as pie-in-the-sky speculation. Conservatives, especially those with strong religious beliefs, were horrified by the thought that our moral failings might be exported to the rest of the universe.

For the historian of technology, one of the most intriguing issues in this area is the fluctuation in expectations of how rapidly progress might be made. The rocket pioneers of the early twentieth century had high hopes, but realized that without government or commercial support their experiments would take a long time to yield fruit. Some visionaries imagined space flight only in the far distant future, while the true skeptics continued to insist that it was impossible. Yet the pulp science fiction authors seem to have felt that it would be only a matter of decades before interplanetary, if not interstellar, flight became a reality. It was the involvement of the military in World War II that allowed the production of the first long-range rockets and opened the way to the space programmes of the 1950s. The ensuing surge of optimism spread far beyond the community of science-fiction fans, with eminent scientists and engineers predicting a moon base by the end of the century, followed shortly by trips to the planets.

The popularity of these exaggerated claims illustrates again some of Gary Westfahl's fallacies of prediction.[2] The assumption that space flight technology would develop rapidly along the lines laid down by the aviation industry ignored both the financial hurdles that would have to be overcome and the sheer hostility of the environment in outer space. Major new technologies would be needed to make systems that could function under these harsh conditions, especially if human beings were to be taken along for the ride. It proved hard enough to develop rocket planes that could operate in the upper atmosphere, let alone rockets that could lift the hardware needed to establish a space station. The facile assumption that adding rockets to aircraft would produce vehicles that would transport passengers from one continent to another at huge speeds helped to drive the expectation that the move into genuine space travel would be an easy extension of the rapid progress in conventional aviation (Figure 8.1).

Fig. 8.1. Anticipation of what an airport would look like in 1985, as depicted at a symposium organized by Trans-World Airlines in 1955. Getty images. The 'aircraft' are clearly rockets similar to those envisioned in science fiction.

The failure to realize just how hostile the conditions would be in space may have stemmed from the still popular belief that life, perhaps even intelligent life, might exist on other planets within our solar system. Mitchell's hope that exploration would soon spread to the planets was encouraged by his belief that even the moon might not be as hostile to life as the astronomers claimed. The science fiction authors who so glibly wrote of Martians and Venusians encouraged the expectation that the planets, at least, would be easy to colonize. This in turn deflected attention from the problems that would be encountered even in Earth orbit.

The Rocket Pioneers

Across America and Europe, small but dedicated bands of enthusiasts began to experiment with rockets, convinced that they were the only means of propulsion capable of reaching outer space. All were amateurs, although many had some technical expertise (if not as retired artillery officers). The influence of writers like Verne and Wells was widely acknowledged and some pioneers contributed to the writing and editing of science fiction. This was no coincidence: the activists were drawn from the community of technical buffs who sustained both popular science writing and science fiction. They formed local and eventually national societies devoted to the design and testing of rockets, and to examining the technical problems that would be encountered in space exploration. Some progress was made in the development of the liquid-fuelled rockets that were soon recognized as the best way forward. But financial restraints limited the groups' effectiveness and it was only when governments and the military became involved in the late 1930s that substantial progress was made.

The numbers directly involved were small – the societies' memberships were numbered in the hundreds – but they did succeed in arousing public interest in the subject. They challenged the skeptics by demolishing popular misconceptions such as the assumption that rockets need something to push against and hence could not operate in a vacuum. More seriously, they found alternatives to technical objections, introducing the concept of the multi-stage rocket to counter the claim that no fuel was powerful enough to achieve escape velocity. They published news of their experiments, sometimes with exaggerated estimates of its potential. This was seen as a vital part of the movement to get space exploration onto the agenda of public debate. Here, the use of science fiction to spread the word was as important as popular science writing and the staging of exhibitions. Only in the later 1930s did concerns about scientific credibility lead to a distancing between the fiction and non-fiction activities, although Arthur C. Clarke remained an exception to this trend. In retrospect, the pioneers were able to convince everyone that they had created the foundations on which later space programmes were built. Whatever the limitations of their technical achievements, their efforts to change public attitudes paved the way for the enthusiasm of the post-war decades.[3]

One of the most widely publicized rocket experimenters was the American engineer Robert Goddard. He had been excited by Wells's *War of the Worlds* as a young man and corresponded with Wells in later life. In 1919, the Smithsonian Institution published his 'A Method of Reaching Extreme Altitudes', in which he advocated sending instruments aloft by rocket, to be returned by parachute. He also suggested that it would be possible to send a rocket to the moon carrying a charge of flash powder that would ignite on impact and be visible from Earth. His paper was publicized in newspapers around the world and in a 1921 *Scientific American* response to criticism Goddard noted that he had received eighteen offers volunteering for a manned rocket ascent, although he had made no such proposal. In 1924, there was another wave of publicity centred on a spurious report that he was planning a manned flight to the moon. In Russia, it was widely assumed that he had actually succeeded. More prosaically, Goddard tested the first liquid-fuelled rocket in 1926. Increasingly annoyed by the publicity, he eventually moved his activities to a more isolated site at Roswell, New Mexico. Here he was supported by the Guggenheim Foundation and from 1929 began to interact with Charles Lindbergh, who had joined the ranks of the space enthusiasts.[4]

Goddard's reluctance to engage with the public was in stark contrast to the attitude of the group which coalesced in New York to form the American Interplanetary Society in 1930. Its first president was David Lasser, a MIT-trained engineer now serving as editor of Gernsback's *Science Wonder Stories*. Lasser also wrote for popular science periodicals and in 1932 he published *The Conquest of Space*, a detailed account of rocket research and future projects with a foreword by the physicist Dr H. H. Sheldon. The book mentioned plans for a transatlantic rocket service which Lasser expected to be operating by the 1950s, but focused mainly on space exploration. Several chapters were devoted to explaining how a manned moon rocket would operate, with descriptions of the conditions the explorers would encounter during the flight and on the moon itself. Thinking further ahead, Lasser noted the possibility of both atomic power and an anti-gravity device. He also suggested that an expedition to Mars would find that the planet supported primitive vegetation.[5]

Lasser was joined by a number of other Gernsback contributors and by G. Edward Pendray, a science writer for the *New York Herald Tribune*, who succeeded him as president. Goddard politely declined an invitation to join the society's advisory board. This was perhaps as well, since the other members were keen to use any available means to advertise their cause both in print and in staging exhibitions. The society was renamed the American Rocket Society in 1934 to deflect attention from the more speculative aspects of its agenda. Rockets were designed and tested in the 1930s, with only limited success – although even the failures were presented to the public as sources of data for scientific research. These efforts were widely reported in the press. Reactions were mixed, but the subject was certainly kept in the public eye. In 1939, the society was associated with Raymond Loewy's design for the Rocketport exhibit in the Chrysler building at the New York World's Fair. This promoted the idea of a transatlantic rocket service, but the imagery was derived from the rockets already becoming familiar in the artwork of science fiction magazines (see colour plate 5).[6]

British rocket enthusiasts faced a major problem: an 1875 Act of Parliament forbade the construction of liquid-fuelled devices. They could, however, engage in the design of rockets and research the problems that would be encountered in space navigation. The British Interplanetary Society was founded in 1930 by the Liverpool engineer Philip E. Cleator, who had been inspired by early exposure to science fiction movies. Unlike its American equivalent, it steadfastly refused to change its name. Given the practical limitations, its remit was always seen as ranging beyond mere rocket research. It began to publish a journal promoting interest in space exploration and in 1937 it moved to London, with the well-known science writer A. M. Low taking over as president. At this point, the young Arthur C. Clarke, a member since 1934, was able to attend meetings and was soon writing for the society's publications.

Low wrote an introduction for Cleator's 1936 book *Rockets through Space*, which paralleled Lasser's *Conquest of Space* by outlining the technical developments needed in rocket design and the problems of space navigation. Like Lasser, Cleator hoped for the eventual development of an anti-gravity drive and predicted that at least primitive life would be found on Mars. The book was not well received by the scientific establishment. As Cleator later noted, the 'persistently antagonistic' journal *Nature* printed a bad review written by Richard Woolley, later much reviled as the Astronomer Royal who dismissed space exploration as 'bilge'. Yet the subject did begin to gain some traction in the British popular science literature. Cleator and Low were among those who wrote on the subject in *Armchair Science*, and the normally conventional *Discovery* included a series of articles on the latest developments and prospects. I. O. Evans included a chapter on the theme in his 1936 book for children, noting the advantages an artificial satellite would have for weather forecasting.[7]

Across Europe, there was a surge of expectation among technophiles during the interwar years. In France, Robert Esnault-Pelterie promoted hopes of space exploration and gained some international recognition though his book *L'Astronautique* of 1930. Soviet Russia experienced a much more active cultural movement inspired by the writings of Konstantin Tsiolkovsky, who had been promoting the topic though both his fictional and popular scientific writings since the end of the previous century. His influence was mostly unofficial and his contributions were only acknowledged by the state shortly before his death in 1935. From 1924, 'cosmic societies' sprang up around the country, driven by a utopian vision of humanity spreading to the stars that was founded more on escapism than technical experimentation. Science fiction novels abounded and some movies were made, with Tsiolkovskii being consulted for the special effects. There was also a hard core of engineering enthusiasts who did test rockets during the 1930s, led by Sergei Korolev. He was purged in 1938, but released for war-work as the Soviet military began to take an interest in solid-fuelled rockets as a supplement to artillery.[8]

It was in Germany that the development of the liquid-fuelled rocket would finally show its potential, although not in a way that most of the experimenters would have preferred. In fact, there were novels imagining future wars with rockets published during the interwar years, although most of the German pioneers were inspired by a more imaginative science-fiction novel. This was Kurd Lasswitz's *Auf zwei Planeten* of 1897, which had depicted a Martian space station above the Earth. Early experimental work with rockets was focused on the Verein für Raumschiffahrt, founded in Breslau (now Wroclaw in Polish Silesia) in 1927 and eventually moved to Berlin. The 'Racketverein' as it was popularly known marked a significant step toward the application of applied science to rocket design, although the full title denotes the founders' ultimate aim of creating a spaceship.

Key figures in the programme included Max Valier, killed by an exploding rocket-car in 1930, Hermann Oberth and Willy Ley. Oberth had proposed using rockets for sending long-distance mail, but his real interest was in space. His 1923 book *Die Racket zu Planetenräumen* was expanded into *Die Wege zur Raumschiffahrt* in 1928, where he proposed a three-stage rocket to reach the moon, with atmospheric braking being used to return the astronauts to Earth. Oberth advised Fritz Lang on the special effects for his 1928 movie *Die Frau in Mond*, ensuring the depiction of realistic-looking rockets landing on the moon. Ley also wrote books promoting the development of interplanetary rockets. His *Die Möglichkeit der Weltraumfahrt* of 1928 cited the work of Dr Walter Hohmann on interplanetary orbits and noted designs by Count von Pirquet for a space station. The construction of such a station was a key goal, seen by these pioneers as the first step in a programme that would ultimately extend to the moon and planets. In 1935, Ley moved to America, where he became a leading advocate of space exploration.[9]

In 1928, the young Wernher von Braun joined the Racketverein and soon became a leading figure in its experiments with liquid-fuelled rockets. But the organization was defunct even before the Nazis came to power in 1933 and von Braun was eventually drawn into a military programme based on the hope of using the rockets as long-range missiles. He moved to the military base at Peenemünde in 1938, and it was here that the V-2 rockets were developed that would eventually be used to bombard England. Von Braun always maintained that his real goal was to create a space programme, the detour into the military being an unfortunate necessity imposed on him by the circumstances. Whatever the truth of these claims, the V-2 finally convinced the whole world that the rocket pioneers were not just impractical dreamers. Von Braun himself went on to become a key player in the American space programme.[10]

Space Operas

Most of the rocket pioneers had been inspired by fictional accounts of space exploration and some had written this kind of material themselves. They did this alongside their popular science writing because they saw it as a valid way of creating a sense of anticipation in the public consciousness. Some of the big names in the golden age of science fiction, most notably Asimov and Clarke, used the combination of fiction and popular science for the same purpose. How influential they were is hard to determine. Young technophiles were certainly enthusiastic about the prospect of space exploration, but the scientific community itself remained unconvinced until the 1940s and the general public must have found it hard to make sense of it all. The message certainly began to spread beyond the pulp science fiction magazines. No one could have missed the Flash Gordon cartoons and movies of the 1930s and the post-war era saw a veritable torrent of what became known as space operas aimed both at adult and juvenile audiences. The millions who panicked at Orson Welles's radio adaptation of *The War of the Worlds* in 1938 must have had at least some awareness of the possibility of space travel, if only by the Martians.

The role played by science fiction (as opposed to popular science writing) was rendered more complex by the wide variety of formats it adopted. The most obviously relevant to the kind of prophecies we are concerned with is the 'hard' science fiction that created its imaginary future by extrapolating from developments already taking place. Adventures could be set in worlds made more realistic by input from those actually engaged in planning the technical advances. There is also a more speculative form of science fiction that goes far beyond what is conceivable in current research to create imaginary futures that may inspire or terrify, but invite ridicule from the uninitiated. Some movie makers used technical consultants to ensure that their space rockets seemed plausible, but Flash Gordon's travels within Ming the Merciless's interplanetary empire were made in rockets that could hardly have inspired much confidence. Writers such as E. E. 'Doc' Smith moved directly to worlds in which interstellar travel was possible. This may have created a sense of wonder among the enthusiasts, but it must have seemed ridiculous to most experts, as well as to the general public. The more visionary futures imagined by later writers such as Clarke did have the power to inspire and helped to bring science fiction into mainstream literary culture – although technophobes such as C. S. Lewis remained hostile.[11]

The authors of both hard science fiction and popular science texts began to make predictions of the timescale within which future developments would take place. They knew that substantial financial support would be needed, but supposed that once the plausibility of the technology was demonstrated, this would be provided by wealthy individuals and commercial enterprises. In the end, it was governments that footed the huge bills involved, driven by military and geopolitical concerns. Once the V-2 had broken the plausibility barrier, most enthusiasts thought it would be only a matter of decades before a space station would be in orbit and bases established on the moon and on Mars. The Apollo programme got the Americans to the moon earlier than even the optimists anticipated, but this turned out to be only a short-term exploration, not a prelude to the much-anticipated move toward permanent occupation of extraterrestrial habitats. Enthusiasm for manned spaceflight waned, and the hopes of those who imagined new technologies offering interstellar – let alone interplanetary – travel remained in the realm of fiction.

The optimism of the pulp science fiction writers was derived from an ideology of individualism and private enterprise. This stressed the role of the inventor or heroic engineer who pioneered a new technology that could be developed by wealthy promoters. A. M. Low, whose popular science writing always stressed the role of the inventor, wrote a space-travel novel in 1937, *Adrift in the Stratosphere*, in which a privately invented rocket balloon is used to explore the solar system. The same image of the individual inventor as pioneer can be found in many early space operas.

This model of progress allowed the author to transit directly from the present to scenarios of cosmic exploration. In 'Doc' Smith's *The Skylark of Space* of 1928, the hero Richard Seaton discovers a new metal which moves at incredible speeds when exposed to an equally mysterious new radiation. Financed by his friend Dick Crane, he builds a spaceship exploiting the discovery that is immediately capable of interstellar flight at speeds greater than that of light. Presumably Smith's training as a chemical engineer had omitted the theory of relativity, but the assumption that some new technology would allow faster-than-light travel would become a stock-in-trade of the more imaginative writers and would eventually enter the mainstream in works such as Asimov's *Foundation* series. Here, the superior technology of the future world is taken for granted, no effort being made to explain how it was developed.

There were many proposals for technologies that would be more efficient than rockets for conventional spaceflight. J. D. Bernal's vision of the future imagined some humans adapting themselves to live permanently on space stations, asteroids and ships setting out on interstellar voyages. He conceded that exploration would begin with rockets, but hoped for exotic new technologies. Perhaps we could control the direction of atomic movements to increase thrust, or develop the ability to transmit electric power from the Earth into space. He also realized that it might be possible to exploit the pressure of the sun's radiation to drive vehicles by 'sails'. Some years later, Clarke became interested in the possibility of discovering a space drive that would modify the force of gravity, mimicking the 'inertia-less drive' imagined by writers such as 'Doc' Smith. Clarke conceded that anti-gravity was impossible, but there were rumours that efforts were being made to achieve this goal in secret government or industrial laboratories. The possibility of amazing new space drives remained commonplace in the juvenile literature of the 1950s, where teenage heroes such as Tom Swift invented vehicles that allowed them to explore the planets. *Tom Swift's Race to the Moon* takes place in his 'repelatron', which simply evades gravity. The popular British author of aviation adventures W. E. Johns moved into science fiction with stories of a Professor Brane who invents a flying saucer driven by the control of cosmic rays.[12]

What made the rapid appearance of these exotic technologies seem plausible, at least to juvenile readers, was the continuing focus on the inventor and his (almost invariably his) ability to innovate beyond the conventions of the scientific community. Individual genius and private finance would open up space, not government-funded big science. This model of innovation was promoted in the interwar years by Low and many other writers, but it survived in the popular imagination. As rockets began to show their real potential in the post-war years, it was still hoped that commercial enterprises might be willing and able to foot the bill in the hope of profit. Here, again, the example of the aviation industry served as the model (conveniently forgetting the role played by military aviation research). Heinlein wrote a number of stories based on this assumption and Clarke's 1951 novel *Prelude to Space* still imagined that the first flight to the moon would be privately financed. By this time, it should have become apparent that governments, driven initially by military interests, were going to be the real driving force, at least in the short term.

The space operas took technological progress for granted, but their optimistic vision of cosmic exploration also drew inspiration from the assumption that life and possibly intelligent life existed throughout the universe, perhaps even on the other planets of our own solar system. By the early twentieth century, it had become widely accepted that the moon was dead and airless, but Mars and Venus seemed more promising. In the decades around 1900, Percival Lowell's efforts to promote his observations of 'canals' on Mars gained wide publicity and inspired Wells to write *The War of the Worlds*. Few could have been unaware of the idea that Mars was inhabited by intelligent beings and the panic inspired by Orson Welles's adaptation of the Wells story in 1938 suggests that it was still current in the public imagination. By this time, the astronomers had pretty well agreed that the canals were illusory and the chances of higher life forms on Mars very slim. Even so, most astronomy texts accepted that there was a good chance of finding primitive vegetation, implying that the planet could easily be colonized. The image of Venus as a hot, swampy world persisted throughout this period, similarly encouraging the hope that life existed beneath its clouds. The fiction writers, of course, found the hope of Martian or Venusian civilizations hard to give up and were still exploiting it as late as the 1950s. Robert Heinlein and Ray Bradbury both helped to keep the idea of intelligent Martians alive long after the scientific evidence had begun to make it implausible.[13]

If ordinary people could be frightened by the prospect of a Martian invasion, they could presumably imagine the human race developing space technology. To what extent they were persuaded to take it seriously is hard to tell. Science fiction and popular science magazines promoted the expectation that rockets to the moon and planets were soon to be developed, but their readerships were restricted to the younger generation. Less specialized magazines sometimes gave a different impression. A *Harper's Magazine* article on 'The Conquest of Outer Space' in 1935 argued that over-enthusiasm of the fiction writers made the whole subject seem too fantastic to be taken seriously – although it did concede that 'the work goes on'. Even some of the most visionary writers of the period did not foresee space travel in the near future. J. B. S. Haldane and Olaf Stapledon predicted humanity moving to the planets only in the distant future when the Earth had become uninhabitable. Haldane's 'The Last Judgment' gave a date of 9,723,841 AD for the first landing on Mars, while it was only the fifth race of Stapledon's *Last and First Men* who moved to Venus hundreds of thousands of years in the future.[14]

Fictional accounts of space travel increasingly promoted the view that a move into space travel was inevitable and their viewpoint did gradually find its way into popular culture. Even Bernal's more serious proposal that humanity might adapt itself to life in space generated press headlines. Movie makers saw the possibilities of science fiction as a model, some productions achieving wide circulation. Lang's *Die Frau im Mond* promoted Oberth's vision of the moon rocket and stills depicting its special effects were widely reproduced in the press. The movie of Wells's *Things to Come* was not a box-office success in America, but it did better in Britain and it too gained significant press coverage. In homage to Jules Verne, Wells had his space pioneers fired from a giant gun, but the advertising posters more plausibly depicted a rocket.[15]

Even boys' comics began to feature adventures in space. Writing in 1940, George Orwell complained about a new generation of comics exploiting fantastic technologies, only to be challenged by Frank Richards (of Billy Bunter fame), who pointed out that such themes had been done to death by the previous generation. A serial on 'The War in Space' had, for instance, appeared in the 1926 *Boys' Magazine* (Fig. 8.2). Comic strips for adults appeared in newspapers and magazines and were soon taken up by other media. The Buck Rogers serial began in 1929, moved to the radio in 1932 and then into film in 1939. Flash Gordon first appeared in 1934 and was syndicated around Europe as well as in America. The first movie was made in 1936 with Buster Crabbe playing Flash in his battle with the evil Emperor Ming. Later movies shifted Ming's capital from the imaginary planet Mongo to Mars, partly in response to the publicity gained by Orson Welles's *War of the Worlds* radio play. The Rocketport display at the New York World's Fair was also designed to convince the public that the technology would soon be available.[16]

The focus on space travel rose to a crescendo in the decades after World War II, by which time the rocket had demonstrated its feasibility. No one could have

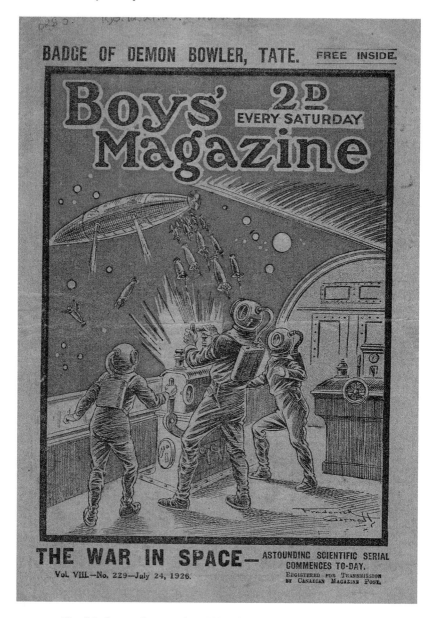

Fig. 8.2. Space adventure in a 1920's boys' comic. Front page of *Boys'*
Magazine, 24 July 1926.

escaped the media's coverage of the theme, whether in the increasingly popular science fiction novels of figures such as Asimov, Heinlein and Clarke, in popular science coverage (where Clarke was again prominent), in the press, on the radio and television, and in film. Heinlein advised on the technical effects in the 1950 movie *Destination Moon*. British radio audiences were enthralled by Charles Chilton's 'Journey into Space' serial (the last radio serial to have an audience greater than those attained by television). Note that Chilton's fictional Jet Morgan is chosen to lead the first expedition to the moon because he is the most experienced pilot on the London to Australia rocket service.[17] When television arrived, viewers were equally enthralled by the exploits of Professor Quatermass.

Juvenile readers were encouraged by books such as the Tom Swift series and by the use of space operas in comics. Both Asimov and Heinlein contributed to the spate of juvenile space exploration stories. British boys of the author's generation were offered the adventures of space pilot Dan Dare in *The Eagle* from 1951. The comic was introduced by its editor Marcus Morris explicitly to counter the influence of horror magazines and Dan Dare echoed the clean-cut image of fictional aviation heroes such as W. E. Johns's 'Biggles'. Johns also wrote juvenile novels about space flight (noted above), as did the popular writer on astronomy Patrick Moore. *Eagle* was soon followed by other comics including a look-alike, *The Rocket*, billed as 'The First Space Age Weekly' and edited by the wartime air hero Douglas Bader. This included a space pilot named Captain Falcon, but also serialized the adventures of Flash Gordon.[18]

Critics who were suspicious of the rising tide of enthusiasm for technology saw the prospect of space travel in a different light. Much of the fictional literature imagined an unreformed humanity spreading out to the stars, extending our sorry record of imperialism, violence and oppression. Those who imagined a new future for humanity often linked this to more rational control of society, and Wells knew that those with moral and artistic concerns would be opposed. The artist Theotocopulos who leads the revolt against the space-gun project in *Things to Come* represents the fear that travel to the planets will cement the rule of soulless technical experts.

For Wells and his supporters, the move outwards from the Earth represented an opportunity for humanity to explore new horizons and gain a higher level of social maturity. The most imaginative writers of the next generation took this insight even more seriously, most obviously in the scenarios created for the human future in Clarke's more visionary novels. Conservative thinkers would not be quieted, however, and in C. S. Lewis they found a champion who would subvert the literary devices of science fiction to challenge the demands of the technocrats. In the three novels sometimes referred to as his 'cosmic trilogy', Lewis adopted the format of the journey to other planets to probe the failings of human nature. In 1938, the first of the series, *Out of the Silent Planet*, imagined a privately financed expedition to Mars using a newly invented drive which exploited 'the less observed properties of solar radiation'. That was perhaps the only convention of the genre that the novel respected. Clarke corresponded with Lewis and even invited him to address the British Interplanetary Society, but the gulf between their viewpoints was too wide for any real interaction.[19]

The Race to the Moon

The deployment of the V-2 ushered in a new phase in the development of rocket technology and laid the foundations for what later became known as the 'space race'. German experts were dispersed among the victorious allied powers and set to work in military establishments created in the expectation that here was a technology that would render the bomber obsolete. Fortunately for the civilian enthusiasts, governments also began to show an interest in using rockets to explore the upper atmosphere and eventually to create artificial satellites. Driven by the same hopes that had inspired research before the war, they also began to fund plans for expeditions to the moon and planets – if only in the hope of demonstrating their technical superiority over rival powers.

The dreams of the science fiction writers now became a reality, and they were cock-a-hoop. Now at last they could pour scorn on the nay-sayers in the scientific community who had insisted that their plans had been illusory. Clarke ridiculed comments made in 1941 by the Canadian astronomer J. W. Campbell, who had rehashed older complaints that no fuel was powerful enough to lift bodies into orbit. The British Astronomer Royal, Sir Richard Woolley, was subsequently much derided for saying as late as 1956 (the year before sputnik) that space travel was 'bilge'. In fact, as Clarke subsequently admitted, the press had misrepresented him – he had been complaining about the exaggerated claims of the enthusiasts and had gone on to warn that putting a man on the moon would cost as much as a major war.[20]

Clarke conceded that Woolley turned out to be right – he and his fellow members of the British Interplanetary Society had vastly underestimated the costs that would be involved. This is why they had imagined that private enterprise, perhaps driven by the hope of extracting minerals from the moon, would be willing and able to finance the projects. They had ignored the possibility that governments might take the research on, seeing no reason why they would fund what would look (almost literally) like blue-sky research. As the real difficulties became apparent, the costs mounted and soon government finance came to be seen as the only way forward. The military was certainly prepared to get involved now that long-distance rockets were a reality. The tensions of the Cold War would drive the development of intercontinental missiles, but they also created a sense of national or ideological rivalry that persuaded the superpowers to use the race into orbit and to the moon as a surrogate arena for demonstrating technical superiority. Emphasizing the peaceful potential for space travel was one way to demonstrate the superiority of their ideology in both industrial performance and cultural potential. The people of the Soviet Union and the United States were most exposed to the propaganda of the space race, but their satellites too were encouraged to lend support. Britain, as a declining superpower, had lingering hopes of creating an independent space programme, soon to be dashed.

The Soviet Union had experienced a wave of enthusiasm for rocket research and space exploration in the pre-war decades. After the war they, like the Americans, acquired some German rocket experts along with captured V-2s, which were tested to evaluate the new technology. Sergei Korolov was rehabilitated and began to play a major role in the programme that eventually led to the development of the Soviet intercontinental missile, the R-7. He was joined by Mikhail Tikhonravov, who was editing Tsiolkovsky's papers and noticed his idea of joining rockets in parallel to create a more powerful vehicle. This technique was used as the basis for the R-7 and most subsequent Soviet vehicles. By now, Tsiolkovskii was being 'canonized' as the pioneer of rocket science, the ninetieth anniversary of his birth being widely celebrated in 1947.

The Soviet Academy of Sciences published the first volume of Tsiolkovskii's edited papers in 1951 and the subsequent years saw a resurgence of technological utopianism in Russian society. Large numbers of science fiction novels were published, many about space travel, along with articles in magazines such as *Youth for the World*, *Technology for Youth* and *Knowledge is Power*. Young people were encouraged in schools and clubs to see themselves as symbols of cosmic progress in which Communist ideology would spread through the cosmos. Plans for space exploration were now announced, with an article by Korolov and Tikhonravov in the *Evening Moscow* paper highlighting the proposal to launch an artificial satellite during the International Geophysical Year, 1957–58. It was this article that was picked up in the United States and stimulated the administration to commence similar plans, in effect starting the space race. The R-7 now became available as a launch vehicle, although the design for the satellite had to be scaled down to ensure success. The launch of Sputnik in 1957 created a worldwide sensation and convinced the United States that it would have to win the race to the moon. Ritchie Calder hailed 'the red moon' as evidence that '[t]he Space Age is definitely here'. Khruschev was impressed and allowed the space programme high priority, while the centenary of Tsiolkovskii's birth was celebrated to symbolize the progress that was expected. The Soviets did indeed have a programme to land instruments on the moon, but this suffered several setbacks and was eventually cancelled in 1974.[21]

In the United States, the campaign to promote space exploration was already underway in the late 1940s. Willy Ley's *Rockets and Space Travel* of 1948 was the first in a series of popular books designed to explain the sequence of development. Ley was soon joined by von Braun, now becoming a leading light in the American rocket programme, and between them they authored books with titles such as *The Conquest of Space, Man on the Moon* and *The Exploration of Mars*. Many were beautifully illustrated by artists such as Chesney Bonestell, who became skilled at depicting imaginary views of scenes in space and on the moon or planets. The sequence of exploration was always the same. First would come the construction of a manned space station, usually a wheel-shaped structure that would spin to mimic gravity for those inside. Expeditions to the moon would then be undertaken, followed by the construction of a permanent moon base in which colonists would live either underground or in pressurized domes. Some illustrations depicted moon rockets that were not the conventional streamlined design, assuming that they would be constructed in space and would land on the airless moon. After the moon, expeditions would go on to Mars and the other planets. The whole package featured as the backdrop to a more visionary image of the human future in the Stanley Kubrick/Arthur C. Clarke movie *2001: A Space Odyssey* in 1968.

There was little emphasis on unmanned artificial satellites in these books. The focus was firmly on manned exploration. This was to be the new frontier that would push American society onwards into a more exciting and more prosperous future. The publicity soon spread to popular magazines and television. A *Collier's* series in March 1952 brought scientists and engineers together to give their views on the claim that 'Man will Conquer Space Soon'. In 1955, Walt Disney's television series included a programme on 'Man in Space' and the subject began to feature in Disneyland attractions. The controversy generated by the launching of Sputnik was, in a sense, a distraction to those promoting manned space exploration, although it was conceded that this was a necessary preliminary to the building of a real space station. The race to the moon began with President Kennedy's announcement in 1961 that the United States would land a man on the moon by the end of the decade. The Apollo project built on the existing public relations apparatus and created a wave of expectation that ended with vast audiences watching the first landing on the moon live on television in July 1969.[22]

In the post-war era, Britain struggled to maintain its status as a world power, desperately trying to keep up with advances in aviation and the new space age technology. Clarke was not the only science fiction author to imagine an independent British programme of manned space exploration – as late as 1959, John Wyndham was still writing about British space stations and moon bases. There were popular books, magazine articles and television programmes predicting the development of interplanetary rockets. *Practical Mechanics* featured spacecraft on its front covers several times in the 1950s. Aviation correspondent Harry Harper published his *Dawn of the Space Age* as early as 1946, while Cleator issued a new book, *Into Space*, in 1953. Patrick Moore, a well-known writer on astronomy, published *The Boys' Book of Space*, with illustrations of space stations and other hardware. He also edited the British Interplanetary Society's new popular magazine, *Spaceflight*, launched in 1956. In 1962, *New Scientist* noted that the Royal Air Force was keen to develop a 'space patroller' for defence purposes, with Barnes Wallis preparing a design using ramjets combined with rockets. Even younger readers were catered for by the inclusion of titles on rockets and space exploration in the popular 'Ladybird' series. All of the popular American books were reissued in London, but their high-quality illustrations only served to drive home the fact that the action was increasingly taking place elsewhere. Britain did develop rockets and even launched a satellite in 1971, but the programme was then wound up – it was the French who increasingly took the lead and played the major role in establishing the European Space Agency in 1975.[23]

Whatever the limitations of the British role, one figure did achieve international recognition in the field. Arthur C. Clarke began publishing actively in the 1940s and soon became both a recognized authority on the space programme, as well as a major contributor to the golden age of science fiction (see colour plate 4). In 1946, his essay 'The Challenge of the Spaceship' had sketched in how the exploration of space might change human values, outlining themes he would develop both in fiction and in commentaries on real-world developments. In the previous year, he had published a paper explaining how artificial satellites in geosynchronous orbits could be used for a global radio and television network, a proposal highlighted in his 1951 book *The Exploration of Space* (Figure 8.3; see also colour plate 4). Through his own television appearances in Britain and the United States, and his collaboration with Stanley Kubrick on *2001: A Space Odyssey*, Clarke went on to become one of the best-known exponents of the belief that humanity's future lay in space.[24]

In the heady days after the war, the proponents of space exploration expected that the development of new rocket technologies would proceed rapidly enough for the first steps to be taken well before the end of the century. Rocket planes for supersonic travel were widely predicted, encouraging the hope that the rapid progress in aviation would lead seamlessly to the development of satellites, space stations and planetary exploration. Clarke and Wyndham both expected moon bases by the end of the century. Heinlein's future history chart predicted a landing on the moon in 1978 and the establishment of a colony in the following decade. Asimov's predictions for the World's Fair of 2014 mentioned the problems of time-lag in radio communications with the moon colony. Those directly involved were equally optimistic: von Braun in 1953 thought that the moon landing would be within twenty-five years.

Plate 11

Drawing by R. A. Smith

SPACESHIPS REFUELLING IN FREE ORBIT

Fig. 8.3. Spaceships refuelling at a space station. From Arthur C. Clarke, *The Exploration of Space* (1951), facing p. 5.

The public, too, were catching the mood of optimism. In 1948, a Gallup poll indicated that only 15 per cent of the US population thought it would be possible to reach the moon. By 1958, 38 per cent thought it would be reached within fifty years and, following the launch of Sputnik in 1957, 41 per cent thought there would be a landing within twenty-five years or perhaps even earlier.[25]

In fact, the first steps took place even more rapidly than the enthusiasts imagined. No one had anticipated the artificial sense of urgency created by the space race between the United States and the Soviet Union, which generated programmes whose main purpose was simply to beat the rival to the goal. The Apollo moon landings created huge excitement, but they were only brief exploratory trips using technology that had no prospect of being extended to allow actual colonization. By the time of the last landing in 1974, the public had lost interest as new areas of concern including the Vietnam War and growing worries about the environment became paramount. Perhaps if there had been a sense that the occupation of the moon would become permanent, there would have been a better chance of keeping up the momentum of popular interest, but NASA was now turning its attention to the space shuttle, unmanned satellites and probes.

As Clarke later admitted, another reason why he and many others had been confident of steady progress was their expectation that rocket technology would soon be making use of atomic power. In 1946, he had written: 'only atomic energy is adequate to lift really large payloads out of Earth's gravitational field' – a comment which he acknowledged made embarrassing reading when reprinted in 1999. Most of the writers active in the immediate post-war decades thought that atomic power – here as in many other areas – would solve all their problems. The failure of the nuclear power industry to develop anything other than reactors far too massive to be used in aircraft or rockets meant that the whole process of space exploration would have to take place using traditional chemical rockets. This, along with a host of other technical problems encountered in the space environment, slowed the progress down to a disappointing crawl.[26]

The late twentieth century thus saw a sequence of developments quite different from that predicted by the enthusiasts. Building on one of Clarke's more practical ideas, artificial satellites have allowed the development of technologies that have transformed many areas of our lives. We do have the International Space Station, although it is a far cry from the magnificent wheeled structures imagined in the 1950s. Unmanned probes to the various planets of the solar system have rekindled public interest in space, but in a form not anticipated by the pioneers. There are still visionary plans for humanity to move out into space, including Gerard O'Neill's designs for space colonies – but these remain in the realms of fantasy until the ability to actually build them in orbit is achieved. Paradoxically, one recent development does return us to a scenario last imagined in the mid twentieth century: the emergence of private entrepreneurs convinced that now at last they can provide an alternative to the government-funded space programmes that have dominated the field.[27]

Plate 1. Poster advertising Alexander Korda's 1936 film of H. G. Wells's *Things to Come*. Alamy images.

Plate 2. Poster advertising the New York World's Fair of 1939. Alamy images.

Plate 3. Futuristic commuter train imagined in an advertisement for the Bohn Aluminum and Brass Corporation in the 1930s. From D. A. Hanks, *American Streamlined Design* (2005), p. 227.

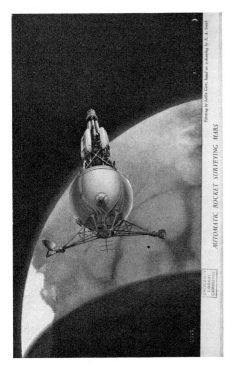

Plate 4. Projected unmanned Mars probe. Frontispiece to Arthur C. Clarke's *The Exploration of Space* (1951).

Plate 5. Imaginary transatlantic rocket, from the front cover of Hugo
Gernsback's magazine *Science and Mechanics*, November 1931. Getty
Images. Note the similarity to the science-fiction rocket images of the period.

Plate 6. Imaginary airport on the roof of a skyscraper, from the front cover of *Meccano Magazine*, May 1932. Author's collection.

Plate 7. 1930s poster advertising the airline KLM. Alamy images. The Flying
Dutchman of legend becomes a reality thanks to the progress of technology.

Plate 8. Poster advertising transatlantic flights by the airship *Hindenburg*,
1936. Alamy images.

Plate 9. Image representing Sir Ronald Ross (discoverer of the malaria parasite) inviting Europeans to colonize the tropics. From Arthur Mee, ed., *Harmsworth Popular Science* (1911), vol. 1, facing p. 233. Author's collection.

9 War

Many projected innovations were hailed by inventors and popular writers as beneficial for daily life and industrial development. But there were always doubters who worried about the destruction of traditional values and ways of life. From suburban sprawl to the inanities of mass entertainment, conservative thinkers could see the downside of each new technology and predicted even worse fates to come. Most worrying was the fact that virtually every new technical innovation, real and imagined, had the potential to be adopted for military purposes. Radio, mechanized transport, aviation and rockets all had obvious warlike applications. The armies, navies and newly formed air-forces of the world soon began to overcome the innate conservatism of the military mindset in the search for new weapons. In some cases, the involvement of the military played a key role in getting a new technology into operation. Worse still was the expectation that new weapons would be created with the power to obliterate whole cities or countries. The public's expectation of benefits from applied science was constantly balanced by fears of a war that might end in global catastrophe.

Some inventors and technical experts were directly involved with the development of new weapons. In late nineteenth-century America, Thomas Edison had already been identified in the public imagination as an inventor who could produce weapons so frightful no nation could resist them. The image of the mad inventor seeking to conquer the world was a staple of popular fiction. As late as 1921, Edison himself predicted that new weapons would make war impossible because no one would dare to use them. Optimists hoped for international control, but few thought that their efforts would succeed. In the Great War, the systematic use of science by the military became more apparent and in the interwar years something like the modern military-industrial complex began to emerge. Everyone suspected that new and better aircraft were being developed, along with more powerful explosives, poison gases and perhaps even biological agents for them to carry. The logic of deterrence was conceived and would remain a staple of national policies into the era of the Cold War.

Some blamed the armaments industry for promoting war – Beverley Nichols's *Cry Havoc!* of 1933 began with an open letter to H. G. Wells calling for a limit on the industry's activities. Others blamed the scientists for allowing themselves to be corrupted by governments and industries. In the 1920s, most academic scientists distanced themselves from war work, but as tensions increased in the 1930s, more became involved – just as they had done during the Great War. A 1931 survey by Victor Lefebure outlined the debates within the scientific community, noting that 'science as a whole is at sea as regards its attitude toward armament and disarmament'. He thought that the inevitable delays that plague the development of any new system might offer the opportunity for international action to control the industry. Some scientists wanted their community to play a role in shaping government policies to limit the potential impact of new weapons. The left-wing scientists active in Britain in the interwar years vilified the system, but when the next war came in 1939 they too rallied to the cause of national defence.[1]

A few optimists thought that the threat posed by new super-weapons was exaggerated. A. M. Low wrote on 'Frightfulness and Humbug', accepting that wars were inevitable in any future society, but insisting that new weapons wouldn't make things any worse. J. B. S. Haldane argued that gas was actually a more humane way of killing people than high explosives and suggested that better weapons might shorten wars and hence lead to fewer deaths. Writing alongside Low in *Armchair Science*, Nichols claimed that scientists could save the world from war by developing weapons too terrible to use. He also suggested using the psychological techniques imagined in *Brave New World* to condition children to avoid weapons. These ideas did little to calm public anxiety, given the barrage of warnings by politicians and military writers about the threat of bombing and a host of novels imagining the catastrophic impact of a future war fought with aircraft, poison gas, germs and even atomic weapons. This kind of literature had already begun to flourish in the late nineteenth century and would rise to a crescendo in the anxious years of the later 1930s. The terrors imagined there were fully as comprehensive as those depicted in fictional accounts of a nuclear holocaust published during the Cold War.[2]

Mechanized Warfare

By the start of the twentieth century, land warfare had been transformed by the rapid-fire rifle and the machine gun. The result was the stalemate of the trenches in the Great War. Motorized transport soon became important behind the lines, but armoured cars proved unable to cope with the rough terrain of no-man's land. There were calls for the development of an armoured vehicle that could cross the trenches and other obstacles – an invention already anticipated in H. G. Wells's 1903 story 'The Land Ironclads'. Here, metal-sheathed vehicles manned by soldiers firing concealed rifles advanced using Mr Diplock's 'pedrails' – large wheels with protruding stumps – to clear obstructions.

When the need for a real-world ironclad became apparent on the Western Front, it was soon realized that the technology to cross rough terrain was already available in the form of the caterpillar tracks introduced for farm tractors, used especially in America. Much controversy surrounds the story of how what became known as the tank was developed by Ernest Swinton and others and how it was deployed with mixed success in battle. But – armed with guns rather than rifles – the machines made enough of an impact to convince forward-looking military officers that they represented the weapon of the future.[3]

Myths surround the development of the tank, most of which have been debunked by later historians. Generals such as Haig did not resist the introduction of the new machines, but did think that there was a need to think carefully about how best to deploy them. In the post-war years, some enthusiasts compared tanks to battleships and imagined fleets of them completely replacing the regular soldiers. Curiously, Wells had predicted that his land ironclads would be accompanied by troops mounted on bicycles. Others conceded that there would still be a role for infantry and perhaps even cavalry to back up the tanks. Lefebure predicted that the infantry-man's rifle 'may in the future be replaced by some super-weapon, say a hybrid between a rifle and a machine-gun'. But it was the tank that would spearhead any attack, and in the 1920s two British military men actively promoted its use to transform land warfare: J. F. C. Fuller and Basil Liddell Hart. Their activities created a further arena for later evaluations, many authorities arguing that their ideas on how tanks should be employed were seriously flawed.

Fuller was a controversial figure with right-wing political sympathies and a taste for the occult. On military matters, however, he favoured materialistic innovation, declaring in his *The Reformation of War* of 1923 that 'invention is an important branch of strategy . . . We must never be content with what we have; without halt we must everlastingly seek for something better.' The traditional soldier was doomed, rendered obsolete by tanks, aircraft and poison gas. Here and in his *On Future Warfare* five years later, he promoted the use of massed fleets of tanks, arguing that they needed to be sealed to protect the crews from gas and could be made capable of submerging in water. Liddell Hart became Fuller's disciple, writing a contribution to the 'Today and Tomorrow' series on the future of warfare. He too promoted mobile warfare led by tanks, and was later led to believe that his ideas had inspired German strategists to develop the technique of the blitzkrieg. The British hoped that by developing better tanks they could limit the need for a large infantry force, but gradually lost the lead during the 1930s. In France, Charles de Gaulle promoted mobile warfare against those who chose the defensive strategy of the Maginot Line. His views were certainly prophetic: when his 1934 book was translated into English in 1940, it was illustrated with photographs of German tanks invading France. The tank went on to play a prominent role in the war and has remained a vital part of military equipment ever since.[4]

The Demise of the Battleship

If tanks were the battleships of land warfare, their naval equivalents were increasingly seen as an outdated technology. Following the launch of HMS *Dreadnought* in 1907, the world's great navies had competed to build ever more of these huge floating fortresses (often called 'dreadnoughts'), with constant improvements being made in guns, armour and engines. The Great War saw only one great naval battle, the somewhat inconclusive encounter between the British and German fleets at Jutland in 1916. After the war, there were further technical developments, eventually limited by international agreements designed to halt what was now perceived as a dangerous arms race. But many forward-looking military thinkers were becoming concerned that the focus on battleships was becoming a distraction. Other technologies, most obviously submarines and aircraft, were emerging to rival the battleships as the key to dominating the oceans.

Submarines played a significant role in the Great War and it was widely assumed (correctly as it turned out) that they would be even more important in any future conflict. But most speculation concentrated on air-power as the most obvious threat to surface vessels. There were a few efforts by novelists to imagine a rather implausible combination, the flying submarine. But as aviation technology advanced, it was clear that both the airship and the aeroplane offered new means of locating and sinking ships. Even before the Great War, *Popular Mechanics* had depicted imaginary conflicts between airships and battleships, echoing a theme already developed by Wells (Figure 9.1). Novels too anticipated future wars in which aircraft made battleships redundant and one, Roy Norton's *The Vanishing Fleets* of 1908, postulated 'radioplanes' – unmanned drones directed by radio.[5]

By the end of the Great War, aviation had developed to the extent that it was obvious (at least to its advocates) that it would dramatically alter the way in which navies sought to control the seaways. In the 1920s, General William Mitchell tried to convince the US Navy that aircraft could sink battleships with bombs – his demonstrations showed that even near misses had the capacity to sink a ship by weakening its hull. His 1925 book *Winged Defense* was widely read around the world, being noted in Britain by naval and aviation commentators such as J. M. Kenworthy and L. E. O. Charlton. Two leading figures in the debate used novels to reinforce their arguments. E. F. Spanner's *Armaments and the Non-Combatant* made a general case for the role of air power, but in his novels *The Broken Trident*, *The Naviators* and *The Harbour of Death*, he depicted Britain being defeated by enemies using air power to destroy both

See p. 165.]   [*Frontispiece*

THE BATTLE OF THE NORTH ATLANTIC.

Fig. 9.1. Airships attacking battleships in the Atlantic. Frontispiece to
H. G. Wells, *The War in the Air* (1908).

fleets and ports. In 1921, the naval architect Hector C. Bywater published a serious study of America's naval rivalry with Japan, later updated to include the increasing role that might be played by aviation. His 1925 novel *The Great Pacific War* imagined a conflict in the early 1930s which included attacks on the American fleet by Japanese aircraft launched from carriers. The possibility that this was read in Japan itself cannot be discounted.[6]

The naval commanders gradually took note of the new developments, although they were slow to concede that the battleship would be rendered completely obsolete. New ships were built and old ones refurbished, with efforts being made to protect the crews against attack by poison gas. Naval officers had a vested interest in minimizing the risk from the air. It was no accident that one of the most vociferous critics of the whole aviation pro-gramme, both civil and military, was the Royal Navy officer (and former submariner) Bernard Ackworth. His criticism of the airship's ability to play a major role had some validity and even the US Navy, which had hoped that airships would make useful scouts, eventually abandoned the technology. The events of World War II would make it obvious that the battleship had no future when faced with aeroplanes armed with bombs or torpedoes. By the time of the Cold War, battleships had become museum pieces.

The Threat of the Bomber

We saw in Chapter 7 that the early enthusiasm for aviation was tempered in Britain by concerns that the country could no longer be protected by the Royal Navy. Aircraft could attack not only troops and ships, but could overfly front lines to attack the enemy at all points. European powers with large capital cities close to potential enemies were especially concerned that their civilian popula-tions could be attacked from the air. The United States was less worried thanks to its relative isolation. The Great War soon showed that the Europeans' fears were justified. London was attacked by Zeppelins, and although these were soon defeated by fighter aircraft firing incendiary bullets, the attacks were resumed by giant aeroplanes. Other allied cities were attacked too, and plans for massive retaliatory strikes on Berlin were in hand at the time of the Armistice in 1918.

In 1916, a British writer, F. W. Lanchester, warned of the increased role that aerial attack would play in the future and argued that science-based improvements in aviation technology were vital for national security. In the following year, Claude Grahame-White and Harry Harper predicted that a future war would start with a surprise attack by air using gas or germs to paralyse major cities. The post-war years did indeed see major developments, reinforcing the claims of those who saw aviation as the most significant area of concern for military planners. Experts such as William Mitchell and Giulio Douhet insisted that command of the air would be essential for victory in any future war. Bombing offensives directed against cities would inevitably be launched when this command was assured. Douhet's *The Command of the Air* emphasized that 'the objective must be destroyed completely in one attack, making further attacks on the same target unnecessary'. Ten planes, he claimed, could carry enough bombs to destroy everything within a half-kilometre circle. The 1933 translation of his book ends with an account of an imaginary future war in which a German victory over France is gained through superior air power.

Fighter planes were improved in the hope that they would be able to repel attacks, but many experts predicted that they would never be able to shoot down enough enemy bombers to prevent massive destruction. Politicians and journalists assumed the worst. The words uttered by British Prime Minister Stanley Baldwin in November 1932 were often repeated: 'The bomber will always get through.' One commentator argued that of all modern technologies: 'Flying is evil through and through, and the reason is not hard to seek. It directly subserves the end of war.' Driven by public concerns, the fear of aerial bombardment played a major role in military and political decision-making and has attracted much attention from historians.[7]

Because many experts thought that defence by fighters was likely to be ineffective, politicians were driven toward the logic of deterrence: the only way to discourage an enemy from bombing one's own cities was to have the capacity to retaliate in force. Some argued that treating civilians as legitimate targets was immoral, but few thought that this would concern a nation determined to destroy its opponents. Calls for international control of military aviation or the setting up of an international air police force were made and ignored. There were concerns that future attacks would be launched without a declaration of war, with the objective of getting in a decisive first-strike. The need to disperse populations and factories to make them less easy targets was discussed and the 1930s saw increasing efforts being made to determine the best ways of protecting the population from bombing.

These moral and social issues were debated against a background of a technical arms race that sharpened fears, but also confused both the experts and the public. Everyone knew that aircraft were becoming bigger and faster. This would allow the production of better bombers, but also encouraged fears that civilian aviation might be misused. It was widely assumed that civilian craft could easily be converted to bombers, a point confirmed by Nevil Shute, who was directly involved with civil aviation. This led the advocates of disarmament to argue that international control would have to be applied to all aviation. There were suggestions for improbable innovations, including aerial torpedoes and fleets of radio-controlled aircraft. A. M. Low predicted balloon forts above cities armed with guns to deter attackers.

Curiously, there seems to have been little interest in the question of how bombers would locate their targets – it was assumed that the attacks would take place in good weather, so cities and ports would be easily visible. Critics such as Ackworth and 'Neon' pointed out the impossibility of accurate aerial navigation in poor visibility. Shute wrote a novel, *What Happened to the Corbetts*, in which German planes attack British cities from cloud cover, having got close to their target flying just above the cloud and using a superior sextant to fix their position before diving into the cloud to begin bombing. In fact, the Germans attacked in 1940 using radio beams to direct their bombers to specific targets. No one except the experts involved imagined a technology such as radar that would allow attackers to be detected as they approached.[8]

Shute's novel was published just before the outbreak of war and predicted an almost complete collapse of civilian life following conventional bombing raids. Many of the 'future war' novels published between the wars imagined the catastrophic effects of aerial attack. In the 1920s, Hugh Addison's *The Battle of London* and F. Britten Austin's *The War-God Walks Again* predicted assaults on Britain by the Soviets. Michael Arlen's *Man's Mortality* of 1933 was one of a number of novels based on the assumption that a shadowy individual or organization would use superior technologies to foment war for their own advantage. In this case, the result was an almost complete collapse of civilization. The Soviets were again the assailants in Leslie Pollard's *Menace* in 1935, but by this time Germany was emerging as the most obvious threat, depicted in Brian Tunstall's *Eagles Restrained* and S. Fowler Wright's *Prelude in Prague*. Like Shute, all of these authors imagined innovations in aviation technology, but saw high-explosive or incendiary bombing as enough to bring a nation, or even a civilization, to its knees. Most novelists – like most aviation experts – thought that the menace would be compounded by the use of horrific new weapons.

Gas and Germs

The best-remembered fictional accounts of air attacks are those in Wells's *Things to Come* and Stapledon's *Last and First Men*, both of which imagine that the bombs would kill by poison gas. Most of the apocalyptic future war novels predicted the use of gas or disease germs dropped by planes to render whole cities or even regions uninhabitable, perhaps for extended periods. The most pessimistic visions described the complete destruction of the warring nations and perhaps even the total collapse of civilization. We tend to think of this level of pessimism as characteristic of the Cold War era, when the threat of a nuclear holocaust seemed very real. But the same level of despair had already emerged in the 1930s thanks to a previous generation of horror weapons.

As Harold Macmillan, later the British Prime Minister, recalled: 'Among other deterrents, expert advice had indicated that bombing of London and the great cities would lead to casualties of the order of hundreds of thousands, or even millions, within a few weeks. We thought of air warfare in 1938 rather as people think of nuclear warfare today.'[9]

There seemed every reason to be pessimistic. Gas had been used in the trenches of the Great War and had a huge effect on the public's image of the horrors of war. Few doubted that the chemists were working on ever-more effective agents, so the assumption that there would be gases that could devastate a whole city did not seem unreasonable. In 1935, Major C. C. Turner cited Professor Gilbert Murray as evidence for the existence of a material of which a teaspoonful could kill a million people. Popular science magazines on both sides of the Atlantic carried articles about the potential dangers. Winston Churchill, languishing out of office, wrote for the popular press emphasizing the dangers posed by gas and predicting that disease germs would also be used as weapons.[10]

A few experts thought the potential horrors of gas warfare were exaggerated. J. B. S. Haldane wrote his *Callinicus* for the 'Today and Tomorrow' series, arguing that the effects of gas were no more horrific than those of high explosive, so a gas war might actually be more humane than a conventional one. Writing in the same series, Liddell Hart echoed this viewpoint: 'Gas may well prove the salvation of civilization from the otherwise inevitable collapse in case of another world war.' A 1934 article in *Armchair Science* cited Dr Freeth of Imperial Chemical Industries, who claimed that popular fears were based on ignorance – gas was an impractical weapon to use on a large scale.[11]

Few were reassured by these opinions and as war loomed in the late 1930s the governments of many European countries began to issue civilians with gas masks. Some idea of the fears influencing public opinion can be gained from the flood of future war novels which rose to a peak in the years before war broke out in 1939. They play a major role in the arguments of later commentators who see the period as one of unmitigated pessimism. Space forbids a detailed description of any but a few of the more effective contributions to the genre. Already in the early 1920s, authors such as William Le Queux and Leslie Beresford (writing as 'Pan') were raising these concerns. Neil Bell's *The Gas War of 1940*, published originally in 1931, was reissued in 1941 – it imagined new gases called 'Halgene' and 'Phosgen', which led to millions dying in torment. The Earl of Halsbury predicted war in 1944, and while noting Haldane's views in a preface, insisted that his own vision of massive casualties from gas was inspired by weapons already existing. Edwin Ramseyer described a gas that rotted flesh off its victims' bones. Ladbroke Black's *The Poison War* predicted a post-war world with a vastly reduced population, but at least imagined a gradual recovery. Others, such as Princess Paul Troubetsky and

C. R. W. Nevinson, worried that the whole Earth might be rendered uninhabitable.[12]

The possibility of biological warfare also played a role in evoking public fears. It featured occasionally in popular science writings and found its way into a few novels. Curiously, these did not necessarily focus on organisms causing human diseases. J. T. Connington predicted worldwide starvation caused by a biological agent that destroyed nitrogen-fixing bacteria, probably the by-product of weapons research. Andrew Marvell foresaw civilization destroyed by an artificial bacterium that coagulated the world's petroleum supplies.[13]

These novels go some way toward vindicating Macmillan's view that there were genuine fears of a global catastrophe resulting from the use of terror weapons long before the nuclear age. There were also warnings about the potential dangers of atomic weapons long before they became a reality (discussed below). In the event, despite the extensive preparations made to defend civilians against poison gas, chemical agents were not used in World War II, although they remain a menace today. Biological weapons also became a reality in the post-war era, although overshadowed in the public imagination by other fears.

Death Rays

Most aviation experts thought that anti-aircraft fire was of little use against bombers. Inventors dreamed of a weapon that would project an energy beam capable of disabling an aircraft (or any other machine) instantaneously at a distance. Perhaps it might also kill or disable people, hence the popular term 'death ray'. The invading Martians of Wells's *War of the Worlds* had used a heat ray to brush aside all opposition. In 1910, Hollis Godfrey's novel *The Man Who Ended War* imagined a scientist who discovers a ray capable of sinking battleships, while Victor Rousseau's *Messiah of the Cylinder* of 1917 had a destructive 'glow ray' (but also 'glow paint' that would protect against it). Ray guns soon became a staple of science fiction, the 'blasters' used in Asimov's *Foundation* series being a typical example. The heat ray would eventually become a reality with the appearance of the industrial laser, famously depicted threatening James Bond in the 1964 film of *Goldfinger*. Between the wars, however, attention focused on radio as a potential means of transmitting enough energy to disable machinery or people. Here, the popularity of the new means of communication created a level of superficial plausibility for the extension of the technology into a military application.

The most widely publicized invention of a death ray was that promoted by Harry Grindell Matthews in 1923–24. Matthews had already gained a reputation for unreliability – he had claimed, among other things, to be the first to transmit a radio message to an aircraft. His new invention used a beam of ultra-violet light to conduct radio energy and was claimed to stop an internal-combustion engine or even explode gunpowder at a distance. The British authorities were not convinced, but the press was enthusiastic. The *Illustrated London News* hailed Matthews's 'electric light ray' as '[a] Wells prophecy that may be fulfilled: an invisible ray for bringing down aeroplanes'. It provided an artist's impression of the ray in action and claimed that for £3 million London could be provided with protection against attack from the air. Popular science magazines were more skeptical, *Conquest* dismissing the claims as moonshine. *Punch* ridiculed the death ray with a cartoon comparing it to the evil eye of antiquity.[14]

Matthews soon dropped out of sight, but popular interest in the idea of a death ray persisted into the next decade. In 1930, Low published a skeptical article in *Armchair Science*, although he subsequently admitted that there was a remote possibility that such a weapon might be produced. Winston Churchill was convinced that German experts were working on the idea and in Britain the Air Ministry again took an interest in the possibility. This was in 1934, when considerable excitement had been generated by a prize of £1,000 offered for a ray that could be demonstrated to kill a sheep at a range of a hundred yards. The inspiration for the development of radar came from Robert Watson Watts's calculation that a radio wave could not transmit enough energy to destroy anything, but might produce an echo strong enough to allow an object such as an aircraft to be detected.[15]

The prize for killing a sheep was never claimed, but novelists were excited by the prospect and several imaginary accounts of future wars included death rays among the exotic weapons that would be used. Stapledon's *Last and First Men* pictured a disintegration ray used to destroy an American air fleet attacking Europe. The machine was then itself destroyed because it was deemed too dangerous – although this did not stop the war and Europe was soon devastated by gas. J. Leslie Mitchell's *Gay Hunter* depicted a world reduced to savagery by war in which the protagonist discovers remains of equipment used to produce heat rays during the conflict. More optimistically, Bernard Newman's *Armoured Doves* and Harry Edmonds's *The Professor's Last Experiment* both imagined the invention of a ray that paralyses machinery and is used by extra-governmental agencies to prevent the outbreak of war. In 1934, Nikola Tesla attracted some press attention with his proposal for what was, in effect, a particle-beam weapon.[16]

The Atomic Bomb

In addition to poison gas, germs and death rays, the possibility that atomic energy might be liberated for destructive purposes was added to the litany of weapons imagined for future conflict. From the earliest investigations of radioactivity, it became apparent that there were vast sources of energy con- tained within the atom which might one day be exploited for good or ill. The possibility that powerful weapons might be developed using atomic energy was popularized by H. G. Wells and others, and there were even speculations that such research might destroy the Earth itself. Atomic energy and atomic bombs featured regularly in the science fiction stories of the 1930s and 1940s, some later predictions coming close enough to what became the Manhattan Project to be monitored by the security agencies. The bomb may have been developed in secret, but its use at the end of the war would hardly have surprised anyone in touch with the speculative literature of the pre-war period.

In the early years of the new century, physicists involved in the study of atomic structure began to speculate in public about the huge amounts of energy that might be available if only they could develop techniques to liberate it. As early as 1903, Frederick Soddy published popular articles hinting at the potential applications of the new science. This was much to the annoyance of Ernest Rutherford, who would long remain skeptical of such predictions. Soddy hoped for a new source of power to benefit mankind, but William Crookes sparked press headlines with a suggestion that the energy contained in a gram of radium could blow the British Navy a thousand feet into the air. Soddy's *Interpretation of Radium* of 1909 concluded with further predictions of the bountiful supplies of energy that might become available, but warned that: 'By a single mistake, the relative positions of Nature and man as servant and master would, as now, become reversed, but with infinitely more disastrous consequences.' It was this book which inspired Wells to write his novel *The World Set Free*, published just before the outbreak of war in 1914. Here, he predicted a future world in which atomic energy becomes so plentiful that it creates economic chaos and precipitates a war fought with atomic bombs. Wells tried to link his ideas with the latest discoveries, imagining explosions that blazed continuously and declined in power with the half-life of the radio- active element employed.[17]

Through the interwar decades, there were occasional flurries of excitement in the press about the possibility of new weapons exploiting atomic energy (although popular science magazines, in Britain at least, carried relatively little on nuclear physics). Concerns that 'splitting the atom' might ignite an explo- sion that would destroy the neighbourhood or perhaps even whole planet, first attributed to Rutherford, were widely reported. The suggestion that the novae or exploding stars observed by astronomers might be the results of such experiments by extraterrestrial civilizations found its way even into the tech- nical literature and was repeated in a French magazine as late as 1940. In 1924, *Punch* depicted a professor appealing to the sporting instincts of his audience to let him go ahead with a demonstration of splitting the atom despite the risk of annihilation. In the same year, Winston Churchill suggested that a bomb no larger than an orange might be produced with the power of a thousand tons of cordite. He repeated similar predictions in 1931 (partly at the suggestion of the physicist Frederick Lindemann) and again in 1938. Churchill would go on to head the wartime administration that approved the first stages in the research on the atomic bomb.[18]

Speculations about an atomic bomb also proliferated in fiction. The play *Wings over Europe*, staged in New York in 1928 and in London four years later, attracted considerable attention with its prediction of atomic bombs used by scientists to force the world's powers to disarm. Harold Nicolson's novel *Public Faces* of 1932 imagined the British testing a bomb underwater in mid-Atlantic – the explosion is far more powerful than expected and the resulting tidal wave devastates the east coast of the United States. John Gloag's *Winter's Youth* depicts an atomic explosive, 'radiant inflammatol', that renders an area uninhabitable for decades. J. B. Priestley attacked the morality of scientists in his *The Doomsday Men*, imagining a group in such despair at the apparent futility of life that they attempt to destroy the world with atomic explosives. Novels by such well-known literary figures ensured that such speculations would reach the general public, but it was in the pulp magazine *Astounding Science Fiction* that stories about atomic weapons proliferated most freely, culminating in Cleve Cartmill's 'Deadline' of 1944, which earned editor John Campbell a visit from the FBI because it so accurately predicted the secret research then underway in the Manhattan Project.[19]

In the year that war broke out, popular science reporting was similarly prophetic. C. P. Snow published an editorial in *Discovery* telling his readers that some experts thought development of an atomic bomb was already under-way. News of Enrico Fermi's plans for achieving fission and their potential implications were widely broadcast in the United States. Such reports were soon suppressed.[20] When news of the atomic bombs dropped on Hiroshima and Nagasaki was released, it generated headlines, but can hardly have come as a complete surprise to anyone who had kept in touch with the speculations of those who knew the potential of nuclear energy. Debate on the social and political implications of the new weapon was limited in the late 1940s by a number of factors. Whole cities had been devastated by conventional bomb-ing in the later stages of the war, and as yet no one appreciated the additional dangers of radioactive fallout. While the United States retained a monopoly on atomic weapons, the threat of all-out nuclear war was postponed. And, in the meantime, every effort was made to convince the public of the benefits that would come from the peaceful uses of nuclear energy (discussed in the next chapter).

Even so, there were some predictions of how fearful a future conflict might become. William Laurence admitted that it might take five to ten years for other powers to develop the bomb, but imagined wars in which hundreds or even thousands of the weapons might be used. General Henry H. Arnold of the US Army Air Force predicted the use of atomic bombs carried by supersonic aircraft, perhaps guided by radio control. He also realized that the rocket technology already developed by the Germans would be married with the new bomb to create an even greater threat, a warning echoed by John Campbell in his factual account of the new technologies. Laurence pointed out that the public as yet did not understand the dangers of radioactive fallout – photographs of a pig swimming in the sea around Bikini Atoll after a test created a false sense of security. Asimov's 1950 novel *Pebble in the Sky* depicted a future Earth treated as a galactic backwater because its inhabitants had adapted to live with an increased level of radiation. He later acknowledged that he was reflecting the common assumption of the time that radiation was not a major health hazard, an assumption that would be challenged by the revelations of the following decade.[21]

The predictions of the pessimists were soon vindicated. The Soviets developed the bomb more rapidly than expected. As the Cold War took hold, each side confronted the other with fleets of jet bombers armed with atomic bombs and the logic of deterrence – which had so evidently failed in the 1930s – re-emerged, along with the spectre of 'mutually assured destruction'. The rockets developed for the space race were merely spin-offs from the projects that created the intercontinental missiles to replace the bomber fleets. More ominously still, the hydrogen bomb, first exploded in 1952, represented a weapon that could completely obliterate even a major city. As if to confirm the worst fears of the pessimists, the H-bomb tests at Bikini finally alerted the public to the true dangers of radioactive fallout. The fate of the fishing boat *Lucky Dragon*, contaminated by a test in 1954, lead to the expectation that a nuclear war might poison whole continents and perhaps lead to the collapse of civilization. If the fears of pre-war doomsayers had seemed far-fetched, the new generation of weapons brought the message home to all. There were disagreements about the true extent of the dangers, but public concern in the West led to increasing demands for a ban on testing nuclear weapons and calls for arms control. By 1960, Herman Kahn's authoritative *On Thermonuclear War* predicted that that by 1975 there might be 50,000 nuclear armed missiles ready to be fired and asked, 'Will the Survivors Envy the Dead?'[22]

Once again, fiction writers led the way in publicizing the deadly effects of the new military technologies. By the late 1940s, science fiction writers were routinely writing of a world either destroyed by a nuclear holocaust, or one in which survivors lingered in savagery plagued by mutants. In 1950, Ray Bradbury's *The Martian Chronicles* (published as *The Silver Locusts* in Britain) brought the message home to a general audience. It concluded with the poignant image of a fully automatic household still functioning despite the extinction of the human race on Earth by a nuclear war in 2026. Humanity survives only via a few colonists on Mars. Novels such as Nevil Shute's *On the Beach* of 1957 (widely serialized in magazines and filmed two years later) drove home the message. Others such as John Wyndham's *The Chrysalids* of 1955 (published as *Re-Birth* in the United States) and Walter M. Miller Jr's *A Canticle for Leibowitz* of 1960 depicted a world in which civilization has been destroyed and the human race corrupted by mutations. Wyndham did at least imagine the possibility of a favourable mutation (conferring telepathic abilities), but saw how those affected would be treated with hostility by a traumatized population.

By the 1960s, the threat of a nuclear holocaust loomed as perhaps the most frightening threat posed by the development of new technologies. The vision itself was not new, but there was now a far more realistic mechanism available to those who imagined future catastrophe. The public image of applied science was seriously compromised. Yet this was but one of a series of revelations that undermined the image of science providing benefits for all. The hope of unlimited power from nuclear energy dissolved as it too was identified as a source of radioactive contamination, while a host of other threats to the environment emerged to reveal the downside of unregulated technological development.

10   Energy and Environment

Many developments predicted in the hope of improving people's lives turned out to have troubling consequences. Apart from military applications, there were less direct challenges leading some to wonder if the benefits outweighed the disadvantages. There were also new vulnerabilities. Even those who relished the new technologies driven by electricity and oil were aware that fossil fuels were not inexhaustible, and pessimists worried that they might soon be in short supply. Other mineral resources had the same limitation, generating real concerns that future progress might be blocked unless research yielded new sources of energy and raw materials. The fear that oil would not last was countered by those who predicted new ways of harnessing what we now call renewable sources of energy. Food shortages – partly arising from the scarcity of essential minerals – might be countered by agricultural improvements or the development of synthetic alternatives. Even so, there were many who worried that science would not be able to solve the problem of food supply, especially as improved medicine was already leading to an increase in the Earth's population.

The search for renewable energy sources became active in the early decades of the century as the use of oil expanded. It soon turned out that there was more oil available than the pessimists feared, leading to a temporary loss of interest in the search for renewables. That search has only picked up more recently as the threat of global warming due to increased carbon dioxide levels in the atmosphere has been recognized. Earlier concerns about possible limits on the supply of oil may have deflected attention from the implications of the 'greenhouse effect' – there was little point in worrying about overheating if there wasn't enough oil to burn.

Other concerns for the environment were already being voiced. An environmentalist movement had begun in the late nineteenth century as the global impact of modern agriculture became apparent. There were calls for the preservation of wilderness and we have noted the growing fears about the effects of urban sprawl as transportation improved. The Dust Bowl of the 1930s revealed the potentially devastating effects of mechanized agriculture and prompted a growing interest in ecology as a science that could address the problem. Julian Huxley joined H. G. Wells to call for better management of the world's resources and to point out that there were often unintended consequences of our meddling with nature. Their concerns anticipated the growing recognition in the 1960s of the harmful effects of chemical fertilizers and insecticides.

One area where far-seeing writers did recognize the possibility of long-term environmental effects was prompted by early efforts to develop renewable energy sources. Schemes for the exploitation of tidal power were proposed, some on a scale that would have had enormous consequences for the local area and perhaps more widely afield. J. B. S. Haldane even worried about slowing down the Earth's rotation, although he was thinking millions of years ahead. Few took such warnings seriously, especially when the expansion of the oil industry seemed to eliminate the need for such schemes. Similar concerns would eventually come to the fore with the search for atomic power. Touted as a vague possibility throughout the early decades of the century, this became a reality in the 1940s. It sparked a wave of enthusiasm that soon faded as it became apparent that small reactors for everyday use were impractical. Nevertheless, the assumption that cheap power from larger reactors would transform the world remained the chief hope of the technophiles. Only in the 1960s did the dangers of radioactive contamination become apparent, as the fallout from weapons tests and industrial accidents was revealed as a major health hazard. Coupled with the growing concerns over other forms of pollution, the changing attitude toward nuclear energy boosted levels of concern over the impact of technology to a new high.

The Crisis of Resources

As the pace of industrialization picked up, there were concerns that the Earth's resources of energy and minerals would ultimately be depleted, undermining all hope of progress. Already by 1865, W. S. Jevons had warned that the global supply of coal would soon be exhausted, especially if industry continued to expand. Jevons's argument was dismissed as alarmist by many, and prospecting around the world revealed that the supply of coal would last for centuries. But it could not last forever, and in the meantime other natural resources were also becoming sources of alarm. In the early twentieth century, attention was increasingly focused on oil as a fuel that was easier to transport and store, and was in fact essential for the new motorized technologies of transportation. Both oil and the minerals needed by industry and agriculture were being used up at alarming rates, generating a sense of crisis that the proponents of technological innovation could not ignore.

In 1898, the physicist Sir William Crookes used his presidential address to the British Association for the Advancement of Science to point out that the world's natural supply of nitrogen fertilizers, essential for modern agriculture, would soon begin to run out. He was able to point to a new scientific development that offered a way out – the artificial fixation of nitrogen from the atmosphere via the cyanamide process. This consumed a vast amount of electricity, but Crookes thought this could be supplied from hydroelectric plants in locations such as Niagara. As it happened, a far more efficient process for synthesizing ammonia was soon discovered by Fritz Haber. Expanded to commercial levels by Karl Bosch, it began production in 1913. Along with mechanization, the Haber process would soon come to be seen as a key resource in the expansion of agriculture.

Crookes's warning served as a wake-up call for others with concerns about the way in which natural energy sources constrained human ambitions. Frederick Soddy may have been inspired by it when he pointed to the advantages that would be gained if the energy of the atom could be tapped in his *Interpretation of Radium* of 1909. By the early 1920s, a number of commentators worried about the expansion of the world's population were harping on the limitations that nature imposed on scientists' efforts to improve agricultural production. A. M. Carr-Saunders and Edward M. East both noted the benefits of mechanization and the breeding of new crop varieties, but insisted that future discoveries would do little to help feed a rapidly increasing world population. In his *The Shadow of the World's Future* of 1928, Sir George Knibbs admitted the huge increase in food production made possible by artificial nitrogen fertilizers and noted that artificial superphosphates had opened up new areas for wheat cultivation in Australia. But he went on to insist that minerals containing essential elements such as phosphorus and potassium were rapidly being exhausted, again suggesting the dangers of unrestrained population expansion. Edward Alsworth Ross made a similar point in response to those who claimed that life would be transformed by the production of synthetic foodstuffs: 'For mankind to multiply carelessly, counting on the early arrival of the synthetic production of foods on a grand scale, would be a pure gamble in human lives.'[1]

Hopes for the production of synthetic foods were not, in any case, fulfilled. Other commentators were more concerned about the limitations on the energy supply. The inevitable exhaustion of fossil fuels was emphasized in Haldane's *Daedalus*, while in 1928 Philip Gibbs appealed to both Soddy and Haldane when he called on scientists to continue the search for atomic power. Attention focused on the availability of oil, with even philosophers such as Bertrand Russell expressing concerns that the supply might be limited. A. M. Low conceded that coal would last for centuries, but argued that this fuel was so dirty that the world was increasingly switching to oil, only to find that this alternative was in much shorter supply. Ritchie Calder's *Birth of the Future* predicted that coal might last 500 years, but oil only a single generation, a point echoed by C. C. Furnas and Percy Lockhart-Mummery. The prospect of civilization itself collapsing as the Earth's resources were used up in the 'Age of Waste' featured in a 1933 science fiction novel by Laurence Manning, *The Man Who Awoke*.

In the late 1940s, American geophysicist M. King Hubbert continued to argue that the supply of oil was limited and went on to develop the idea of 'peak oil' according to which supplies would soon begin to decline. By the 1960s, his views were being vigorously contested by the oil industry, increasingly confident of its ability to discover and exploit more reserves. The situation changed to such an extent that in 1964 an expert from the Shell oil company could look ahead to a 'Plenitude of Petroleum', encapsulating developments that had rendered the earlier concerns irrelevant, at least for the time being.[2]

The Search for Renewable Energy

The early estimates of the world's oil resources underestimated what would be discovered by further exploration, but in the short term there was a boom in the search for alternative sources of energy. Most of the renewable sources of energy we take seriously today were suggested and some attracted considerable publicity, but none was taken up on a large scale. As the oil boom continued unabated, there was progressively less incentive to invest in research into technologies that would require major innovation to become viable. Only in the later twentieth century would the anticipation of a looming environmental catastrophe renew interest in these projects.

In 1924, *Conquest* reported a discussion of alternative power sources that took place at the British Empire Exhibition held at Wembley. These included geothermal, tidal, solar and wind power. The lack of progress over the following decades is illustrated by the fact that in 1940 Waldemar Kaempffert's chapter entitled 'After Coal – What?' in his *Science Today and Tomorrow* listed the same future possibilities. The only newcomer was atomic energy, then just beginning to emerge as a serious contender. Both surveys missed one alternative to petroleum: the production of alcohol from fermenting vegetable material. Even before the Great War, the *Harmsworth Popular Science* serial suggested that alcohol from fermented potatoes would become a popular alternative to petroleum fuels, noting that schemes to develop the technology had been proposed in Germany. Progress was slow, however, and in 1936 C. C. Furnas was still suggesting this kind of biofuel as a future development. He noted that huge amounts of land would be taken up to grow the necessary crops and called for research to develop an artificial equivalent to fermentation using catalysts to produce suitable chemicals from carbon dioxide.[3]

Geothermal energy was another alternative that attracted only limited attention. At the turn of the century, Charles Parsons had proposed boring a hole

Fig. 10.1. Experimental solar power plant constructed in Egypt in 1913. From Harry Golding, ed., *The Wonder Book of Engineering Wonders* (no date, c. 1920), p. 230.

twelve miles deep into the earth to continue his research on the formation of diamonds. He noted that an immense amount of heat could thus be brought to the surface, but since there was no realistic prospect of boring this deep, the idea remained dormant. Camille Flammarion suggested a similar scheme in which water would be pumped down to be turned into steam, but the *Harmsworth Popular Science* dismissed it as impractical because of the huge size of the hole required. There were places where volcanic steam was available closer to the surface and in 1911 a working power station was built at Larderello in Italy. The possibility of using this as a model for schemes elsewhere was noted by J. N. Leonard and by Lockhart-Mummery, the latter warning that it would only become practical in the long term. Larderello remained the only functioning geothermal plant into the 1950s.[4]

Another alternative with geographical limitations – at least with the available technology – was solar power. The only plausible way of harnessing this in the early twentieth century was by concentrating the sun's rays to generate steam for turbines, which seemed to make sense only in the tropics. In 1913, the British established an experimental system in Egypt, reported later by Calder in his *The Birth of the Future* and for younger readers in *Meccano Magazine* and a popular survey of engineering developments (Fig. 10.1). Low wrote of huge mirrors being built to concentrate the sun's rays in the tropics and noted that a system using thermocouples to produce power was being tested in California. Solar power was also mentioned in Leonard's *Tools of Tomorrow* and Furnas's *The Next Hundred Years*, but both conceded that real progress would only be made if the photovoltaic effect could be harnessed to develop electricity directly. In 1938, Dr Godfrey Cabot endowed the Massachusetts

Institute of Technology with funds to be devoted to the study of solar power sources. By the early 1950s, the main project was still based on the use of the sun's heat to generate low-pressure steam for turbine. Photoelectric cells were also being developed, although conversion rates of only a few percent would delay their commercial application for some time.[5]

Most accounts mentioned the necessity for developing better ways of storing electrical energy if intermittent sources such as the sun or wind were to become viable. French engineers made detailed studies of wind energy and designed turbines for use in exposed areas, but they too felt the need to develop new storage systems. Most English-language commentators were dubious of wind-power's potential to solve the anticipated power shortages when fossil fuels ran out, although the possibility of harnessing the wind did gain some support. The mega-cities depicted in Wells's 'A Story of the Days to Come' and *The Sleeper Awakes* had giant windmills built on their roofs. Haldane's *Daedalus* also opted for wind to replace fossil fuels, on the grounds that it was unlikely that atomic power would become available.[6]

The source of renewable energy that attracted most attention was that derived from water, both from inland rivers and from the tides. The proposals were often for schemes of such a magnitude that there would inevitably be massive consequences for the local environment. The infrastructure would also be hugely expensive to build. During the interwar years, schemes to control the flow of rivers, both for water management and for power generation, were built at several sites in North America. The potential for significant impact on the environment was immediately apparent, as illustrated by the title of a 1921 article in *Conquest*: 'Is Niagara Doomed?' The author foresaw development on such a scale that the whole flow of the river would be utilized for power, leaving the famous waterfalls dry. *Tit-Bits* cited an unnamed authority who pointed to Niagara as evidence that hydro-electric power might replace coal. *Conquest* thought hydro-electric power could soon replace coal completely in Scotland. In the 1930s, *Discovery* hailed proposals to build dams at Victoria Falls in Africa and across South America.[7]

Plans to make use of tidal energy were also proposed, usually involving barrages built across river estuaries or bays with relatively closed mouths. The Bay of Fundy was a popular target because of its enormous tides, but there were also proposals for Passamaquoddy Bay in Maine, the Severn estuary in Britain and the bay of Mont Saint Michel in France. Lockhart-Mummery mentioned the most ambitious proposal of all, damming the Straits of Gibraltar and lowering the sea level of the Mediterranean. In 1933, Evans showed his juvenile readers an artist's impression of a futuristic installation which rose and fell with the tides. Most authorities conceded that these schemes would only be built in the distant future when the need to replace fossil fuels became critical. When he wrote 'The Last Judgement', Haldane switched his attention from wind to tidal power and imagined a future where the human race became dependent on this source over a period of millions of years. Eventually, he predicted, so much energy would have been derived that the Earth's rotation would be slowed down to the extent that the planet became uninhabitable. No one else was thinking this far ahead, and most of the schemes have remained unrealized.[8]

Nuclear Power

Following the discovery of radioactivity, reports began to circulate of an entirely new source of energy that might be derived from the atom. Many physicists were skeptical, but there was always enough support to keep hopes alive. Popular magazines and newspapers lapped up reports of the potentially unlimited power that might be obtained almost for free if only a way could be found to unlock the energy of the atomic nucleus. They also warned of the potential military applications, but the hope of solving the impending crisis of energy supply was enough to focus attention on this area of science. When the atom was finally 'split' in 1932, there were headlines that reinforced the scientists' suspicion of press sensationalism. The first report of the successful release of nuclear energy provoked further alarm, but as the world tried to come to terms with the atomic bomb, the scientific and technical establishments realized that promoting the peaceful uses of atomic energy was a valuable way to allay public concern.

Before the start of the new century, Gustav Le Bon in France had attracted press attention by claiming that the discovery of radioactivity showed the interchangeability of matter and energy. Le Bon was soon discredited, but key figures in the emerging field of atomic physics soon took up the theme. The most prominent was Frederick Soddy, who gained attention as early as 1906 with talks promising that radioactive elements could yield enormous amounts of energy. Soddy was aware of the argument that fossil fuels would eventually be exhausted and saw atomic energy as the only prospect of freeing humanity from the limits imposed by conventional technologies. His *Interpretation of Radium* of 1909 expounded this message in detail and inspired Wells to write his *The World Set Free*, in which the discovery of how to generate cheap nuclear energy creates economic chaos and war, leading to the eventual emergence of a rationally ordered society. From this point on, the prospect that science might solve the riddle of how to tap this new source of energy would be a continuing source of press speculation.[9]

The scientific community was by no means united in encouraging these hopes. Ernest Rutherford confirmed that in theory a huge amount of energy was available, but soon emerged as a leading skeptic of claims that science might find a way of exploiting it. When the atom was eventually 'split' in 1932,

Rutherford broadcast on the BBC arguing that the hope of generating energy by this means was illusory and he was reported in the *New York Times* as saying that the whole idea was 'moonshine'. Robert Millikan, originally an enthusiastic researcher in the field, changed his mind and similarly dismissed hopes of practical nuclear power. The contribution to the 'Today and Tomorrow' series on physics didn't even mention the study of the atom.[10]

Other skeptics included Haldane, whose *Daedalus* held out no hope in this direction. The same cautious attitude was expressed in Calder's *The Birth of the Future* and Lockhart-Mummery's *After Us*. As late as 1940, Kaempffert – originally an enthusiast – suggested that atomic energy was unlikely to be available soon. Low mentioned atomic energy only as a vague possibility. Furnas noted that the potential benefits were enormous, but thought that the research needed to unlock the secret of atomic power would take vast amounts of time and money, a view also expressed in the magazine *Discovery* in 1934. Furnas, with tongue in cheek, advised against buying shares in any company proposing to develop atomic power. Leonard was more positive about the prospect of further research, but shared Wells's concern that unlimited cheap power might cause economic chaos. He also expressed concerns about radiation dangers and hoped the research would not succeed in his own lifetime.

Optimists included Birkenhead, who was sure atomic energy would be available by 2030, while Gibbs waxed lyrical on the prospects: 'If the scientists can get hold of it, liberate and utilize that atomic force – some of them think they are getting close to the secret – mankind will be put into possession of power so illimitable that all previous forms of energy such as coal and oil and water will become negligible and man himself will be the master of the very source and origin of power.'[11]

For the popular press, the prospect of unlimited power was almost as newsworthy as the threat of being blown up by atomic explosions. Soddy's views were echoed almost immediately in the *Harmsworth Popular Science*, which wrote of 'Power beyond our dreams' that would 'cost almost nothing'. In 1920, *Tit-Bits* quoted Sir Oliver Lodge on the power that might become available thanks to research currently going on. Four years later, the *Illustrated London News* described the work of physicist T. F. Wall under the title 'Seeking to Disrupt the Atom: Immeasurable Energy'. *Tit-Bits* returned to the theme in 1925 to emphasize the huge amounts of energy that could be obtained from '1 oz. of electrons'. In 1932, the artificial disintegration of an atomic nucleus was finally achieved by J. D. Cockcroft and Ernest Walton, prompting lurid newspaper headlines about both the power available and the threat of explosions. Even the popular science magazines could not be trusted – *Armchair Science* cited Sir James Jeans to the effect that with atomic power a piece of coal the size of a pea could drive the *Mauritania* across the Atlantic and back. This was just the kind of exaggeration that Rutherford worried about, and he took steps to get more balanced accounts into the better-quality magazines. The science writer J. G. Crowther's article contained only a brief reference to atomic power and insisted that it lay in the distant future.[12]

Science fiction writers took it for granted that atomic energy would be the power source of the future. J. J. Connington's *Nordenholt's Million* of 1923 imagined atomic energy being developed by the survivors of a global catastrophe. The best-known example of the theme is Robert Heinlein's 'Blowups Happen' of 1940, which pointed to the potential hazards of using a power source that might accidentally turn into a bomb. In the following decade, articles predicting the potential benefits of nuclear power began to appear regularly in American magazines. When news of the atomic bombs dropped on Japan was released, it was immediately coupled with articles expressing the hope that the same principles would soon be applied for peaceful purposes. In Britain, the *News Chronicle*'s story of the Hiroshima bomb ended with the hope that instead of creating havoc, the new technology 'may become a perennial fountain of world prosperity' and was followed by an article by Sir John Anderson arguing that the next step was to control the force.[13]

Thus began a campaign on both sides of the Atlantic to focus public attention on the potential benefits rather than the danger posed by the atomic bomb. Optimistic stories of abundant power and of miniature atomic reactors driving cars and aircraft abounded through the late 1940s and the 1950s. In 1953, President Eisenhower gave a widely reported speech pointing to the benefits of nuclear power. Two years later, the first International Conference on the Peaceful Uses of Atomic Energy met in Geneva and heard confident predictions that the new power source would mean an end to the threat of a global energy crisis. In 1954, Lewis Strauss, chair of the US Atomic Energy Commission, claimed that nuclear power would become too cheap to meter. In Britain, there was an 'Atom Train' exhibition in the late 1940s, followed by 'Atoms for Peace' (touring by bus) in 1955. Nuclear power was presented as the obvious way forward at the 1951 Festival of Britain. As late as 1969, Nigel Calder – responding to a growing tide of concern over the harmful effects of applied science – hailed nuclear power as one of twelve recent science 'booms' and repeated the claim that it freed the world from the threat of an energy shortage. By this time, the term 'nuclear energy' was being promoted in the hope that it would deflect attention away from any link to the atomic bomb.

Calder was aware that the public mood had now begun to change. From the start, there were some experts who realized that the optimistic predictions of small-scale reactors powering even automobiles were unlikely to be realized. Engineers focused on the basic principles of reactor design simply assumed that further research would solve the problem of miniaturization, not realizing that the need to shield against radiation would limit how far the process could be taken. Already by 1947, John W. Campbell was suggesting that within ten years the hopes for nuclear-powered gadgets would be dashed. He was even skeptical over the general expectation that nuclear energy would become widely available: 'For the general public the much-advertised atomic age is about to open with a dull thud of disappointment and a growing conviction that it has been badly oversold.' Other experts were skeptical about the prospect of atomic cars, but more hopeful that large-scale power generation would be practical. George Gamow still thought that atomic planes might be possible, but conceded that the future lay with 'giant central power plants', with the energy distributed by atomic storage batteries. P. M. S. Blackett noted that many physicists worried about the nuclear industry's connections with the military. He argued that the most appropriate use of atomic power would be in countries such as Australia, where the population was widely dispersed and there were few sources of fossil fuels. Like several other commentators, he pointed out that the United States had abundant supplies of oil and coal, but would probably develop atomic energy for strategic reasons. Even so, the Atomic Energy Commission reported to Congress that it would be twenty years before significant quantities of power became available from nuclear sources.[14]

Plans for nuclear power stations were actively developed and promoted in the 1950s. A 1956 survey by physicist E. W. Titterton noted that Britain had commissioned its first station, Calder Hall at Windscale in Cumbria, and hoped to generate 15 per cent of its power from atomic reactors within ten years and nearly half by 1975. A detailed account of the Calder Hall project by Kenneth Jay predicted that twelve stations would be built over the next ten years and compared the current design to the Model T Ford – a great step forward, but only an indication of better things to come. Others were not so sure. Harrison Brown's *The Challenge of Man's Future* of 1954 suggested that it was by no means certain that the nuclear option was the best way forward and expressed concern over how radioactive waste would be handled. Public alarm over the dangers of radioactivity in the environment grew as the effects of fallout from bomb tests became apparent. Calder Hall delivered its first power to the national grid in 1956, but in October of the following year a fire at the plant caused significant contamination of the surrounding area.[15]

Governments and the nuclear industry continued to talk up the prospects, but the whole project was beginning to unravel as the public became more aware of the environmental consequences. Campaigns for nuclear disarmament allied themselves with the growing environmentalist movement, seeing the by-products of the nuclear age as one of the more serious areas of pollution. In 1962, veteran science writer Ritchie Calder published his *Living with the Atom*, a response to what the nuclear industry regarded as the overreaction of the popular press on both sides of the Atlantic. He admitted that the press had created an atmosphere of fear, but went into considerable detail on the genuine problems that were emerging. If the proposals for new nuclear power stations went ahead, by the year 2000 the industry would be generating a ton of radioactive waste every day. Disposal would become a real problem and Calder mentioned a number of options, including vitrification and storage in salt mines. He also raised the possibility of hydraulic fracturing of rock strata so that liquid waste could be pumped into the earth – a suggestion that resonates oddly with our modern debates over the consequences of fracking.[16]

Changing the World

One of the more bizarre suggestions for the peaceful use of atomic energy was the possibility that it could be used to transform the Earth's geography and climate. Atomic bombs could be used to excavate canals and restructure coastal geomorphologies. The vast amounts of power that would become available could be used to melt the polar icecaps and to desalinate enough seawater to create fertile land in the Sahara. In the initial wave of enthusiasm, the fact that melting the icecaps would raise the sea level by a hundred feet was passed over as something to worry about later. All these ideas were proposed by the eminent biologist Julian Huxley, newly appointed head of UNESCO, at a packed meeting in Madison Square Gardens in late 1945. He had been encouraged to think along these lines in discussions with Desmond Bernal. Nothing could more clearly illustrate the scientists' failure to appreciate the dangers of radiation. Edward Teller was still advocating the use of 'clean' H-bombs to remove arctic ice in the 1960s, but by this time, such ambitious plans were viewed with increasing skepticism.[17]

Huxley's enthusiasm also points us toward a different issue. He was a biologist associated with nature conservancy and the study of wildlife. That he could so blithely suggest massive interference with the Earth's climate requires us to re-examine the relationship between the science of ecology and the rise of environmentalism. In fact, Huxley, a one-time supporter of H. G. Wells, was firmly committed to the ideology of progress through the rationally planned application of science. He certainly played a role in promoting ecology and was concerned to preserve at least some of the world's wildlife against the encroachments of humanity. But the study of natural interactions was not seen as part of a programme to limit the applications of technology – it was to be used to anticipate and control the impact those applications would have on the world. Unfortunately, those consequences would turn out to be far more serious than anyone anticipated.

Melting the icecaps became conceivable only with the inflated expectation of unlimited nuclear energy. But the hope that the Earth itself could be transformed for the benefit of humanity had been promoted since Wells himself made his first predictions. Fiction writers had speculated about inventors who could alter weather patterns at will, and by the 1940s efforts were being made to produce rain and control storms. Mechanized farming and the transplantation of crop species around the world were transforming landscapes and whole ecosystems. Chemists expected to solve the problem of fertilizers. Biologists hoped that genetics would allow the breeding of crops and animals to solve the world's food problems. Even if the promised synthetic foods did not materialize, scientific agriculture would deal with the population problem. If earlier transformations had had unanticipated consequences, ecology would allow us to understand natural relationships and plan more carefully in the future. As yet, hardly anyone recognized the possibility that the global climate might be warmed by the by-products of burning oil.

As ever, there were those who feared the hubris of the scientists and the planners. From the Industrial Revolution onwards, the pollution surrounding production facilities had become obvious and urban overcrowding was the source of huge distress. As modern facilities expanded for industry and home life, conservatives regretted the passing of traditional ways of life. Complaints about urban sprawl and ribbon development accompanied the rise of new transportation networks and city developments. Moves to protect areas of wilderness or traditionally farmed landscapes began in the late nineteenth century as the threat posed by mechanized agriculture intensified. Some thought that transforming the landscape might have beneficial effects on the climate. The Dust Bowl of the American west in the 1930s demonstrated that drastic changes brought about by intensive farming made a region vulnerable to climatic fluctuations we had no power to control. It took time for the message to sink in, but by the 1960s the new threats posed by atomic radiation and the increased use of pesticides produced a change in the public mood that forced even the enthusiasts to tread more carefully. Conservation was no longer a matter of preserving a few areas of wilderness, and ecology became a matter of global interest.

The plans to use atomic energy to reshape the climate hoped to realize a vision that had long enthralled those who saw science as a means to transform the world. In the early decades of the century, futuristic novels imagined inventors who used amazing devices to control the weather or even to alter the inclination of the Earth's axis of rotation. As with the future-war novels, the end product was usually economic conflict and chaos. On a smaller scale, Sir Oliver Lodge and others saw the possibility of clearing fog by electrostatic precipitation. There was a popular belief that radio broadcasting had affected the weather. Writing for young readers in 1933, Evans dismissed this as nonsense, but went on to suggest that high-tension beams or rockets could be used to break up clouds. Nothing came of these ideas and it was only in the 1940s that the first serious efforts were made to control the weather. Irving Langmuir promoted cloud-seeding as a means of producing rain, generating both headlines and cartoons in the popular press. A striking image showing a scientist operating a weather control machine was used to illustrate a *Collier's* article on the topic in May 1954. Soon the US military took an interest and in the following decade cloud-seeding became a weapon of war in Vietnam, with controversial results. Weather control for peaceful purposes gradually faded from the agenda, at least until revived in response to growing concerns about the global climate at the turn of the century.[18]

Controlling Life

Weather control would have brought obvious benefits to agriculture, but this was an area already being transformed by mechanization and the breeding of new crop varieties. In the early twentieth century, life scientists began to argue that their increased ability to understand the breeding process offered new ways to control nature for human betterment. New technologies were thought to offer the prospect of 'speeding up evolution' under human control. For Hugo De Vries, founder of the hugely influential 'mutation theory', the study of variation and heredity in a controlled environment would allow us to direct evolution along channels of our own choosing. In his *Species and Varieties*, originally published in 1904, he argued that 'if it should once become possible to bring plants to mutate at our will and perhaps even in arbitrarily chosen directions, there is no limit to the power we may finally hope to gain over nature'. Perhaps the threat of food shortages could be avoided by the development of much more productive crops.[19]

There were excited reports in the press that technologies derived from the physical sciences would be applied to generate the new varieties that crop and animal breeders needed. Daniel MacDougal was one of several botanists who thought that treatment by radium might allow mutations to be produced at will. Following Herman Muller's 1927 demonstration that X-rays also produced an increased mutation rate, there were efforts by the breeding firms to use the technique to provide new varieties that might turn out to be useful. In the following decade, chemicals such as colchicines were also used in the hope of producing similar results. Despite the high expectations, few varieties of any real use were discovered. In the rush of enthusiasm to promote the peaceful uses of atomic energy in the post-war decades, irradiated seeds were widely touted as providing farmers and even gardeners with new crops or flowers.[20]

Despite wide public interest in these proposals, American breeders were more sympathetic to the less theory-driven techniques of Luther Burbank. De Vries's large-scale transformations were not, in fact, true mutations as understood by the emerging science of genetics. But genetics itself fed upon the hope that by helping breeders to control the flow of hereditary material within a species, it would speed up the creation of more productive crops. In Britain, senior figures such as William Bateson and R. A. Fisher were associated with horticultural research institutes and the American Breeders' Association took an early interest in Mendelism. Perhaps the improvement of traditional breeding methods might have more to offer than the high-tech applications of the new physics.

As the century progressed, most commentators agreed that breeding practices informed by genetics were having a beneficial effect on productivity, but few were convinced that the results would be revolutionary. Sir John Russell told a BBC audience in 1937 that there might be better cabbages in the future, but they would still be cabbages. Pessimists feared that improvements in yields would be insufficient to stave off the threat of widespread starvation (see Chapter 11 below). Bolder thinkers looked to synthetics, or perhaps to hydroponic farming in areas unsuited to conventional agriculture. There were equally bizarre ideas derived from conventional hybridization – in 1929, *Armchair Science* hailed the 'cattalo', a cross between cattle and the buffalo, as a future source of meat. A few years later, Furnas pointed to the weaknesses of purebred animals and plants in a chapter warning that agricultural technology was not keeping up with population increase.[21]

Soviet Russia at first sought a different path, as the agricultural expert T. D. Lysenko convinced Stalin that he could improve yields by environmental modification of the crop species. Genetics was vilified as a capitalist science and many geneticists were purged. There were ambitious claims that the new techniques introduced by Lysenko and his followers would allow the country to eliminate its chronic food shortages. Little was realized in practice and Russia missed out on the advances made in the West, leading to claims that the whole episode showed the dangers of political involvement with science. More recent studies suggest that Lysenko's work may have dealt with effects now recognized by the science of epigenetics.[22]

Furnas's *The Next Hundred Years* also devoted a whole chapter to the problem of pest control, noting the evolution of insect varieties resistant to the then-common arsenical insecticides. The breeding of new crop varieties did pay off in the end, however. In the post-war world, new strains of rice and wheat staved off mass starvation in what became known as the Green Revolution. Development of new insecticides also proceeded apace, but resulted eventually in the massive over-use of DDT in the 1950s. The Green Revolution itself depended in part on increased use of pesticides in the developing world. The resulting crisis for wildlife highlighted in Rachel Carson's *Silent Spring* of 1962 was a key element, along with growing fear of nuclear radiation, in the emergence of an environmentalist movement critical of scientific interference with nature. Even so, Carson's arguments may have deflected attention from the hazards posed by other chemicals such as organophosphates that could be used as substitutes. The hope that humankind could control the production of food through applied science did not die easily.[23]

The Science of Balance?

Concerns over the global impact of applied science certainly intensified in the 1960s, but some of the issues had been recognized much earlier. The change in public attitudes was partly due to the increasing scale of the problems, but it also reflected growing suspicion of the ideology of planning for progress. In the interwar years, supporters of Wells and the other advocates of a rationally planned future could still argue that science would provide the tools to manage problems as they emerged. There was a new science available – ecology – that offered methods for studying and predicting the interactions between species and the environment. By the 1960s, it would be seen as the obvious handmaiden of environmentalism, but for Huxley and the other progressives, it was merely a new tool for predicting and managing the consequences of technological expansion.

Threats to the environment from industrialization were obvious to all, and there was an increasing awareness that the expansion of mechanized agriculture was impacting on the world's wildlife. That deliberate hunting could drive species to extinction had become obvious, and concerns were expressed over the fate of whales as hunting continued unchecked. There was also a growing willingness to admit that there might be unintended consequences of human activities. In a chapter entitled 'The Price of Progress', Furnas noted a number of examples where introduced species had wreaked havoc on native wildlife. He also pointed out that well-meaning efforts to protect species in national parks could have harmful consequences: culling mountain lions had led to the proliferation of deer and the destruction of natural vegetation patterns. There was also growing awareness that intensive farming had an effect on the local climate. The settlers who opened up the prairies of the American west believed that 'rain follows the plough', a view echoed in an *Armchair Science* article in 1933. The same article also noted that deforestation caused a decrease in local rainfall, a view endorsed by environmentalists anxious to protect or renew natural forests.

The year 1933 is significant because it saw the start of the environmental catastrophe that became known as the 'Dust Bowl'. Mechanized farming had destroyed the prairie grassland, but had been profitable only when rainfall was adequate. When the rains failed, as they often did in the Midwest, the unprotected earth simply blew away in giant dust storms. The 1936 movie *The Plow that Broke the Plains* blamed mechanization for the disaster, but in the previous year ecologist Paul B. Sears had published his *Deserts on the March* criticizing the failure to appreciate the inevitability of climatic fluctuations. He challenged the over-optimism of the farmers who had assumed that modern technology would always be able to conquer nature, a belief 'at once touching and dangerous'. Geographer Carl Sauer pointed the finger squarely at the attitude that encouraged reckless exploitation in the hope of profit:

We are too much impressed by the large achievements of applied science. It suits our thinking to rely on a continuing adequacy on the part of the technician to meet our demands for production of goods. Our ideology is that of an indefinitely expanding universe, for we are the children of frontiersmen. We are prone to think of an ever complex world created for our benefit, by anthropocentric habits of thinking.[24]

As an ecologist, Sears represented what we would now call the environmentalist wing of a new science devoted to the study of natural interactions. He used examples from around the world to illustrate how human interference with the natural balance of species in an area resulted in disaster. But ecology had not emerged as the handmaiden of environmentalism and it did not reflect a unified vision of how nature should be studied. Some ecologists did adopt a holistic viewpoint, treating natural communities almost as superorganisms with a life of their own. For grassland ecologist Frederick Clements, each region had a natural 'climax' vegetation which could be restored if disturbed. But even Clements was forced to accept that once the topsoil had blown away from the prairies the damage might be permanent. He joined other ecologists urging the US Soil Conservation Service to recommend grazing as a less damaging way of exploiting the dryer areas.

Other ecologists adopted a more mechanistic view of nature which depicted species as competing with one another to occupy territory. This in turn allowed them to associate with precisely the managerial attitude that Sauer had blamed for the Dust Bowl. Ecology helped us to understand the complexity of natural relationships, but this did not mean that we should simply abandon efforts to manipulate those relationships to our advantage. On the contrary, by acquiring a better understanding of nature's complexity, we would for the first time have the ability to interfere in a way that foresaw potential damage and minimized its effects. In the conclusion to the 1947 reissue of his *Deserts on the March*, Sears praised H. G. Wells for calling ecology the 'science of prophecy' – although Wells's own view of humankind's relationship with nature was very much at odds with the environmentalist position that sought to minimize the disruption caused by agriculture.

The managerial approach to nature was clearly displayed in the collaboration between Wells and Julian Huxley that produced their best-selling survey *The Science of Life* in 1931. Huxley had made important studies of animal behaviour in the wild and was by no means blind to calls for the protection of at least some areas of natural beauty. He consulted widely with respected British ecologists. But he was a firm supporter of Wells's programme for reshaping the world along rationally planned lines. Scientists and other experts were needed to guide the process. The current system was wasteful and risked damaging the Earth precisely because industrial and agricultural entrepreneurs paid no attention to the long-term consequences of their actions. *The Science of Life* listed the many disasters for wildlife that had been produced by human interference with nature and insisted that these harmful consequences could be minimized by better understanding. It presented the need to ensure balanced communities as an extension of the Wellsian economic programme, with ecology playing a key role by informing managerial decisions to rule out unintended consequences. The chapter entitled 'Life under Control' also included a call for the human population to be limited. In the words of historian Peder Anker, Wells and Huxley saw themselves as 'The Board of Directors in the Economy of Nature'.[25]

The tension between the technological utopians and the environmentalists would reverberate through the rest of the century. As Sears reported in the 1947 edition of his book, important steps were now being taken to minimize disruption in the Midwest – yet there were more Dust Bowl episodes in the following decades. Water from aquifers was exploited to make farming sustainable, without thought for how long the underground supplies would last. In 1948, Fairfield Osborn's *Our Plundered Planet* highlighted the threats posed by overexploitation around the world and reiterated the argument that the Earth was not a gadget we could tinker with at will. Conservation, not exploitation, was the key and science must be weaned from its assumption that it can offer us unlimited ability to control nature.

Huxley's call for the use of atomic bombs to reshape the Earth indicates the continued expectation that planned technological expansion could produce an artificial environment that was better for all. The US Atomic Energy Commission became a leading source of funds for ecological research in the hope of managing the effects of fallout. Yet by the 1960s, the cumulative effect of various environmental catastrophes had begun to generate a new level of public awareness of the potential dangers of unrestrained exploitation. Between them, the threats posed by fallout and chemicals began to gain a hold on the public's awareness. The loss of wildlife highlighted in *Silent Spring* reinforced the concerns of environmentalists and reshaped the popular perception of ecology as a science that was devoted to the study and protection of wild nature.

An area of concern that has now become a source of controversy was just starting to rear its head in the 1960s: global warming. It had been known since the nineteenth century that carbon dioxide was what we now call a 'greenhouse gas', helping to trap the sun's heat in the Earth's environment. Swedish physicist Svante Arrhenius predicted that increasing the level of carbon dioxide could produce a significant warming across the whole planet. Soon there were suspicions that the climate was indeed warming up. In 1938, engineer Stewart

Callendar provided the Royal Meteorological Society with evidence of a warming trend and identified increased levels of carbon dioxide as the most likely cause. Most authorities dismissed the idea of a major impact on the environment or thought that the effects would be beneficial. Few anticipated that consumption of oil would rise to the extent that its by-products could affect the whole Earth, and in the post-war decades hopes for atomic-powered transport would have deflected attention from the issue. A warmer Earth might enjoy increased agricultural production and in 1954 Harrison Brown even advocated artificially increasing the level of carbon dioxide to promote plant growth. In the following decade, the first stirrings of concern over this new source of pollution began to circulate, although they were still very much in the background. In 1964, Dr Roger Revelle of the University of California noted that the Earth did seem to be warming from increased carbon dioxide levels, but also predicted advantages from using the waste heat from nuclear power stations to warm the oceans.[26]

11   Human Nature

Haldane's *Daedalus* proclaimed that it would be biology, not the physical sciences, that would take the lead in transforming our lives. There was already plenty of evidence to back up the claim. Breeders were producing a range of new plant and animal varieties, while ecology was starting to understand the impact of human activities on the environment. More worrying was the prospect that evolution or an artificial equivalent might impact directly on the human species. Wells's *Time Machine* imagined a degeneration of the race in the distant future, a theme echoed in Stapledon's *Last and First Men*. It was widely supposed that we might ultimately lose superfluous structures such as teeth and nails. Of more immediate concern was the rising tide of efforts to control human nature artificially through the application of biological science – this was what Haldane was really drawing attention to. In the early twentieth century, scientific medicine was getting into its stride, generating enormous hopes for future cures and a plethora of wildly exaggerated claims. Breeding techniques originally applied to animals and plants could be applied to the human population, perhaps leading to a race of supermen. There were even hopes that science would soon produce entirely artificial life forms.

Traditionalists worried about the potential misuse of the techniques involved. The fear of germ warfare pointed to the darker side of efforts to manufacture living things. The spectre of Mary Shelley's *Frankenstein* haunted the imagination of those who feared the arrogance of the scientists' hope of controlling human life. The robots in Čapek's *R.U.R.* were biological entities, not mechanical, and a 1939 novel by Philip Chadwick imagined artificially manufactured soldiers used in war.

The hope that living organisms, humans included, could be manipulated like machines derived from the advances in experimental biology pioneered by figures such as Jacques Loeb. The new medical science could thus be identified in part with the materialist philosophy that had begun to challenge traditional religion in the Victorian era and remained active in the new century. Yet modified versions of a more spiritual way of thinking persisted and we should not assume that the materialists had it all their own way. Alexis Carrell achieved international recognition with his 1935 book *Man the Unknown*.

Although his reputation derived from his work as an experimental biologist, Carrell advocated a holistic view of life that sought to integrate the latest research with a more spiritual dimension.[1]

For critics, however, the darker implications of materialism formed a genuine threat to traditional values. In *The Island of Doctor Moreau*, Wells had pointed to the potential horrors that might be unleashed by the heartless use of the latest vivisection techniques. Moreau tried to turn animals into humans, but more worrying still was the arrogance of those who sought to apply physiology, biochemistry and psychology to control the abilities and behaviour of individual human beings. Whether through short-sighted benevolence as in *Brave New World* or deliberate cruelty as in *1984*, the potential for a nightmare future seemed all too real. The eugenic movement sought to transform human nature at a different level through artificial selection, but liberal thinkers feared its focus on restricting the reproduction of the 'unfit', often by repressive means. In this case, the potential for horror was only revealed by the atrocities of the Nazi regime in Germany, although the same attitude was reflected in a less severe form in the eugenic regimes of several other countries.

The issues raised by the prospect of biological control straddle the long-standing debate over the roles played by nature and nurture in the determination of human character. On the side of nurture, the experimental biologists and psychologists who thought they could condition behaviour or enhance performance dealt with individual development. Any effect they produced would have to be duplicated in every individual and every generation to have any permanent influence on the race. The geneticists who believed that inbuilt nature predetermined by heredity held the key dismissed individual modifications as trivial. They operated with whole populations, determining which characters would be transmitted to future generations and thus duplicating the effects of evolution on the species as a whole.

In practice, the distinction between the two schools of thought was seldom so extreme because both approaches were often used in tandem. The 'Today and Tomorrow' series contained a title written by Ronald Campbell MacFie entitled *Metanthropos: Or the Body of the Future*. Like Carrell, MacFie was a well-known opponent of materialist thought, yet he acknowledged the possibility of future changes in the body and accepted that breeding could have some effect, while stressing the need for better environments and improved medical care. What became known as 'reform eugenics' called for improvements in the environment and education, as well as less intrusive methods of ensuring that the least fit individuals would not breed. The well-conditioned citizens in *Brave New World* were produced by manipulating individual development, not heredity – although Huxley also supported eugenics during the 1930s. Appeals to nature or to nurture, to coercion or persuasion, all were entwined in a complex mix of scientific and ideologically loaded efforts to shape the human future.

The Conquest of Disease

Like many of his contemporaries, MacFie believed that medicine would soon be able to cure most diseases and appealed to growing knowledge of hormones as the key to improving human nature. Scientific medicine – medical practice informed by the latest biological discoveries – had begun to influence the field in the late nineteenth century and was now the most visible sign of progress. A generation of young enthusiasts was told about the past triumphs of medical science and encouraged to participate in future progress (even if the medical profession was sometimes a little slow to admit the need for change). Several areas were identified as the most promising ways forward. Bacteriology and later the study of viruses had begun to offer means of controlling some of the more serious infectious diseases, and it was assumed that future progress would eliminate such threats altogether. Better understanding of diet, including the role of vitamins, offered real hope of eliminating diseases caused by malnutrition. There was also increased recognition of the harmful effects of many substances used in industry and daily life. All of these areas provided information that might allow direct control of particular diseases, often leading to exaggerated hopes in press reports.

For obvious reasons, public attention tended to focus on the benefits that would affect developed societies. But there was another dimension to the rise of medical science when it was applied on a global scale. In the previous century, the ability to control tropical diseases had allowed Europeans to penetrate areas hitherto off-limits for health reasons. This imperialist dimension continued in the early decades of the new century. In the years before the Great War, the *Harmsworth Popular Science* serial hailed the work of Sir Ronald Ross in aiding the conquest of malaria with an image labelled 'Can science colonize the tropics?' (see colour plate 9). The role of medical science in allowing the Panama Canal to be built was also noted – earlier efforts had been defeated by the ravages of fever. The same point was made by commentators on the population problem such as Edward Alsworth Ross. While hailing the steady improvement in health in the developed world, they worried that as the medical benefits spread to local populations in the tropics, the declining death rates would not be matched by corresponding decreases in birth rates (Figure 11.1).[2]

Back home, the new developments focused attention on public health, and it was here that some of the most effective (if less spectacular) advances were being made. Some commentators who favoured eugenics also conceded that progress could only happen if everyone lived in a decent environment. The *Harmsworth Popular Science* included sections on eugenics written by Caleb Saleeby, an obstetrician who campaigned tirelessly for better public health (he even called for the establishment of a national health service).

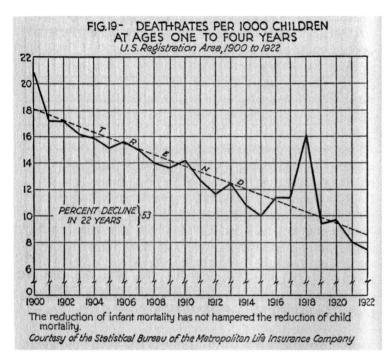

Fig. 11.1. Graph showing declining death rates in infants, from Edward Alsworth Ross, *Standing Room Only* (1927), p. 90. The downward trend is clearly indicated (despite the spike caused by the influenza pandemic of 1918), but Ross worried about the population problem that would result if the same medical benefits were extended to the developing world.

The only medical chapter in H. G. Wells and Julian Huxley's *The Science of Life* dealt almost exclusively with the promotion of a healthier lifestyle. Later in the 1930s, Ritchie Calder included a chapter on health care in his *The Birth of the Future* and followed it up with a whole book, *The Conquest of Suffering*, which included an introduction by Haldane. While noting triumphs over particular diseases, he focused mainly on the need for better living conditions and the elimination of harmful materials produced by industry. Dr Edward Mellanby, the discoverer of vitamin D, was quoted predicting the eradication of most diseases in the near future and complaining about the limited funding for the Medical Research Council. America, too, came in for criticism:

If newspapers devoted a tenth as much space to the great and successful drive against infant mortality in the United States as they do to gangsters, we might realize that when it comes to killing, the virus of measles makes Al Capone look like a professional life saver. There ought to be a squad of a hundred bacteriological detectives hunting down the criminal.[3]

The problem was that spectacular promises of new cures attracted all the headlines. In the early decades of the century, both radium and X-rays were seen as new technologies capable of wiping out diseases. In 1921, *Conquest* published a photograph of an X-ray machine under the title 'A Cure for Cancer?' Radium and X-rays had their uses, of course, but hopes for the rapid eradication of diseases were soon dashed when it became clear that in fact their effects on healthy tissues were harmful. The discoveries of vitamins and hormones were also hailed as offering potential cures for most diseases in magazines and futurological books by A. M. Low, C. C. Furnas, Philip Gibbs and (for younger readers) I. O. Evans. *Conquest* called vitamins 'the mighty atoms of the diet' and thought that their lack might be the cause of cancer. Low also published occasionally on health in *Armchair Science*, a 1935 article under the title 'The Health Guard Will Advance' predicting the cure of most physical diseases and even insanity.[4]

Hopes that cancer might be cured by X-rays or vitamins remind us that then as now this disease was of particular concern. Three books with the title *The Conquest of Cancer* appeared during the first half of the century, suggesting that the rise of scientific medicine was at last raising hopes in this area. In 1907, Saleeby argued that surgery had been shown to be ineffective and insisted that discoveries by a Dr Beard pointed to pancreatic ferments as the cause. If only the medical authorities would listen to the proponents of this theory, a cure would soon emerge. A contribution to the 'Today and Tomorrow' series by H. W. S. Wright pointed to irritations such as pipe smoking as the cause and called for regular screening so that cancers would be detected at an early stage when cures by surgery or X-rays could be effected. Fear of going to the doctor was the principal danger, since it left cancers to develop to a stage when there was no hope of a cure. The same point was made over twenty years later in 1947 by George Bankoff, who justified his 'optimistic and flamboyant title' by arguing that enough progress had been made to allay the traditional fear of visiting the doctor. Progress remained slow, however, and when Ritchie Calder returned to the theme in his *The Life Savers* of 1962, he listed cancer in his final chapter on 'Unfinished Business', arguing that it had now been shown to be a complex disease with at least some genetic components.

The Elixir of Life

The slow progress on cancer exposed the absurdity of the exaggerated claims for miracle cures which hit the headlines throughout this period. Of these, the most persistent were those centred on the discovery of the hormones secreted by the ductless glands. As endocrinologists made progress in studying these substances and their effects, enthusiasts advertised procedures that would cure impotence, rejuvenate the whole body and perhaps prolong life itself. In the 1920s, their claims hit the headlines around the world, attracting the attention of the rich and famous and inspiring numerous literary figures and satirists. The controversy over the work of Serge Voronoff and Eugen Steinach has continued to interest historians in part because it throws light on the cultural anxieties of the time created by changes in the roles traditionally assigned to the sexes.

There had been claims that rejuvenation through biochemical techniques might be possible in the late nineteenth century, but these had been rejected by the experts. In the 1920s, suspicion was rapidly eroded by a raft of new discoveries. Julian Huxley attracted headlines when the newspapers got wind of an article he published in the journal *Nature* describing the effects of thyroid extract on the axolotl, a salamander that retains its gills into adulthood. The extract caused the creatures to metamorphose into normal salamanders, which the *Daily Mail* interpreted as offering control over sex and development and the prospect of eternal youth. Huxley tried to stem the tide of over-enthusiasm, discovering as he did so that he had a talent for popular writing about science. In December of that year, *Conquest* carried an article on hormones as chemical messengers and proclaimed that, despite the press sensationalism, science was 'on the threshold of discoveries of great benefit to medicine'. Later in the decade, the 'Today and Tomorrow' title on the future of chemistry suggested that it would only be a matter of time before hormones could be synthesized, allowing us to banish every kind of abnormality: 'Then the path will be open to the perfecting of the race.' As late as 1936, when the spate of enthusiasm had waned, Furnas could still write that despite all the hokum the hormones 'lord it over our lives'.[5]

Serge Voronoff was a Russian emigré working in Paris in the years after the Great War. He had a successful medical career and became independently wealthy through marriage before he began his experiments transplanting testicular material from apes into human males in the hope of rejuvenating them and perhaps of prolonging life. When he began to claim success, the press took an interest, and in the 1920s he became a global celebrity (Figure 11.2). The wealthy and famous queued up to have the operation and he began to claim that he might be able to cure cancer too. He also tried to create superior breeds of farm animals, especially sheep. Eugen Steinach worked in Vienna and developed a very different technique using bilateral vasectomy with the aim of diverting some of the seminal fluid into the body where it might have rejuvenating effects. He too attracted worldwide attention and the support of eminent figures who claimed to have benefitted from his operation. By 1923, there were over 100 surgeons in New York City alone offering the treatment,

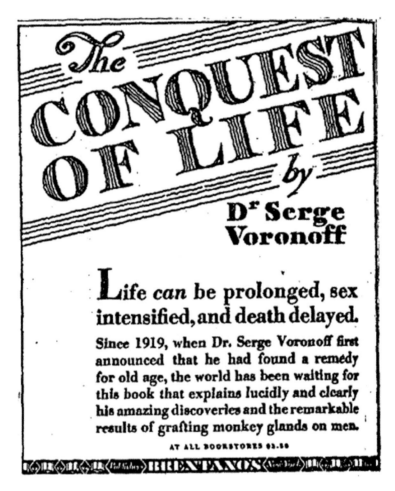

Fig. 11.2. Poster advertising Serge Voronoff's book on rejuvenation. Chicago
Tribune archives.

and charlatans such as John R. Brinkley continued to use it long after it had
been discredited.

These techniques appealed to the prevailing obsession with the glands as
determinants of personality and reflected a complex mix of anxieties arising
from the rise of feminism and fears of racial decline. Voronoff was featured in
the *New York Times* under the title 'Science Promises an Amazing Future'.
Magazines such as *Scientific American* also praised his work and in Britain
Armchair Science devoted several articles to the topic, including favourable

comments from Low. The poet W. B. Yeats claimed to have benefitted from the Steinach operation. Novelists and cartoonists celebrated, lampooned and criticized the public's obsession with sex, glands and vitality. Nor was the ferment confined to the English-speaking world – Germany experienced a surge of enthusiasm for 'rejuvenation biology', eventually purged from the racial hygiene movement by the Nazis. In Russia, the Soviet authorities condoned wild speculations about further applications of biochemical experimentation.[6]

The expectations of the patients derived from their own personal insecurities, but to their scientific proponents these techniques had far-reaching consequences for the human race. In *The Conquest of Life* of 1928, Voronoff wrote of the potential for grafting to boost the potential of children and suggested that the results would have permanent effects equivalent to his production of new varieties of sheep: 'Why not try creating a race of super-men, endowed with physical and intellectual attributes very superior to ours?' He appealed to mothers to entrust their children to him so they could 'contribute to a new chapter in the history of humanity'. This was evolution without Darwinian natural selection, depending on the alternative mechanism (still widely accepted at the time) of the inheritance of acquired characteristics. Enhanced abilities imparted to the individual, in this case, by grafting, would be passed on to future generations and shape evolution.

A leading exponent of this non-Darwinian theory, often known as Lamarckism, was the biologist Paul Kammerer, whose experiments with the midwife toad would eventually be discredited by the geneticists. Kammerer monitored Steinach's work in Vienna and in a 1924 book insisted that the rejuvenation work arose from the theory of evolution 'to which it is closely related'. He too hoped for permanent improvements in the human race. Another supporter, Louis Berman, attacked the geneticists for their claim that all characters are rigidly predetermined by heredity. In his book *The Glands Regulating Personality* of 1930, he promoted the ability of the biologists to enhance performance and ended with a chapter on the future of human evolution. Here, he insisted that the effects of the glands must somehow become involved with the genes so they could become embedded in the race. The biochemists would be able to control mutation and thus direct the development of the race toward Utopia.[7]

Others took a more pessimistic view. In 1926, Julian Huxley published a science fiction story 'The Tissue-Culture King', in which an endocrinologist helps the chief of an African tribe to control his people with hormones. Two years later, Gibbs's *The Day After Tomorrow* noted the enthusiasm for glandular therapies, but warned that the technique might be adopted by governments seeking to control their populations. Aldous Huxley's *Brave New World* became the best-known warning of how the new biology could be used to produce a race of willing slaves, although he imagined even more drastic ways of interfering with individual development. Other novelists commented on the problems that might be encountered by individuals who opted for rejuvenation. Techniques involving ovarian grafts were developed for women and Marie Corelli's *The Young Diana* described a woman who achieved immortality and was transcended onto a new mental plane. Gertrude Atherton's *Black Oxen* of 1923 imagined women being rejuvenated as well as men, but warned of social disruption caused by tensions between the genuinely young and the rejuvenated, and the increased risk of overpopulation. A technical study by Peter Schmidt praised the beneficial effects of the Steinach operation, but pointed out that the annuity and life-insurance industries would be ruined by mass rejuvenation.[8]

Critics also worried that the grafting of animal tissue might ultimately result in dehumanization. Lord Dunsany's play *Lord Adrian* ended with its protagonist being shot for encouraging the animals of his estate to rise against humanity. Aldous Huxley's *After Many a Summer* ended with another nobleman being revealed at the age of 200, although he had degenerated into a state resembling an ape. Having accidentally discovered the secret (raw fish guts) in the early nineteenth century, he had now matured into the original state from which humanity had emerged by evolutionary foetalization. Aldous Huxley's fictional Earl of Gonister had been transformed by exactly the same process as Julian's axolotles.

By the time Huxley's novel was published in 1939, the craze had petered out as it became increasingly obvious that grafts from other species were always rejected. Alexis Carrel, an expert on transplantation, had warned of this from the start, but had been widely misrepresented. The British endocrinologist Swale Vincent argued that the whole field would be brought into disrepute by the exaggerated claims touted in the popular press. Even so, as late as 1937, a BBC talk asked 'Do You Want to Live Longer?' and hinted at the existence of a 'life gland'. Ten years later, Waldemar Kaempffert asked 'Why Can't We Live Forever?' and pointed out that if we could, the future evolution of the human species would be blocked. Steady progress was actually made in some areas, resulting in techniques such as hormone replacement therapy. Yet, in the 1960s, the pattern of the Voronoff affair was repeated as Paul Niehans of Geneva offered rejuvenation based on thyroid extracts to the rich and famous, including Pope Pius XII. He also claimed to 'cure' homosexuality with extracts from testicular cells.[9]

The Decline of the Family

The debate over rejuvenation fed into wider concerns over changing patterns of sexuality and men's fears of women's emancipation. There were also fears that scientific interference with reproduction would undermine the traditional structure of the family. Better contraception would inevitably lead to sex being divorced from reproduction, but the real enthusiasts for the new biology predicted that things would go much further. Haldane's idea of ectogenesis (completely artificial fertilization and embryological development) would not only make sex purely recreational, it would give whoever controlled the process immense power to shape the future of humanity. Where the supporters of eugenics held that the race could only be modified by restructuring its gene pool through artificial selection, the new experimental techniques would manipulate the environment in which individuals would be conceived and raised. In principle, this approach offered even greater control, creating individuals shaped for a preconceived purpose and conditioned to enjoy their state. The enthusiasts no longer wanted to perfect the traditional human form – they sought to define what the future form of the race would be.

Two of the boldest projections – Haldane's *Daedalus* and Bernal's *The World, the Flesh and the Devil* – were published in the 'Today and Tomorrow' series. It also included Bertrand Russell's *Icarus*, which anticipated some of the fears articulated more forcefully in *Brave New World*. The series included a number of other works commenting on issues related to reproduction. They appeared in the context of a debate on the changing role of women prompted not only by the availability of better contraception, but also by changes in how the household would be run.

Commentators and novelists hailed or derided the prospect of a world of sexual equality. Russell's wife Dora contributed to the 'Today and Tomorrow' series, stressing the need for better-spaced pregnancies, and told men that there would only be peace between the sexes when they understood what women really wanted. Vera Brittain took a science-fiction approach in her text, predicting that the effect of better sex education and the freeing of sex from reproduction would be a world not of promiscuity, but of 'enlightened monogamy'. She quoted a fictional professor of the future thus: 'The scientific achievements of the late twentieth century liberated humanity for all time from the age-old enslavement to animal instinct, physical limitation, and biological change.' Anthony Ludovici used the series to attack feminism, suggesting that women would take over society, reduce men to only a tiny percentage of the population, and stifle originality. F. C. S. Schiller opted for eugenics precisely because he did not believe that the work of the endocrinologists would yield viable methods of controlling personality. Herbert Spencer Jennings took the opposite view of eugenics, but conceded that if we could develop a method of 'uniparental reproduction', it would be easier to plan the future of the race. Like Bertrand Russell, he feared that this would allow the ruling class to perpetuate its view of society.[10]

There were many other efforts to predict what might happen. Low was a supporter of feminism who thought that women would soon (literally) be wearing the trousers. He urged them to turn their hand to inventing, exploiting their different perspective on what was really needed. Marriages would be for a limited time only, and children would be looked after outside the traditional family. Psychologist John B. Watson predicted a world in which promiscuity would be normal in adolescence, after which couples would settle down to stable relationships. But couples would only raise their children if they chose to – otherwise they could be sent to 'wonderful' facilities where they would enjoy the best possible upbringing.[11]

The prospect of children being raised in state-controlled institutions filled the critics with apprehension. It was bad enough that here their mental characteristics might be shaped by the new hormonal technologies and by applications of Watson's own behaviourist psychology, but if the children were produced entirely artificially, the state could exercise control from the moment of conception. This was the nightmare future that Huxley would predict in *Brave New World*, although in fact these fears were already being debated years before that book appeared in 1932.

Although Haldane's *Daedalus* brought the idea of ectogenesis openly into the public debate, his speculations had begun to circulate several years before his book was published in 1924. Haldane and the Huxleys interacted socially from 1919 and Aldous Huxley's first novel, *Chrome Yellow* of 1921, has a character who expounds ideas very similar to those on which *Brave New World* would be based:

In vast state incubators, rows upon rows of gravid bottles will supply the world with the population it requires. The family system will disappear; society, sapped to its very base, will have to find new foundations; and Eros, beautifully and irresponsibly free, will flit like a gay butterfly from flower to flower through a sunlit world.

The sunlit world will, however, include biologically distinct tribes defined by psychological conditioning to work and obedience and going through life in a 'rosy state of intoxication'. This was already putting a negative spin on the ideas that became the basis for *Daedalus*. In the book, Haldane acknowledged the role of heredity, but saw the intrusive role of biology applied equally through ectogenesis (which he predicted would first be achieved in 1951), allowing the chemical control of development to produce the characters and emotions the state desires. Artificial reproduction would allow the implementation of a rational plan of development for mankind, a view he promoted in an article for the London *Evening Standard* in 1927.[12]

For Huxley and others who appreciated traditional values, these predictions could more easily be read as the road to disaster. The caricature in *Chrome Yellow* was echoed more soberly in *Icarus*, Bertrand Russell's contribution to the 'Today and Tomorrow' series, and in his *The Scientific Outlook* of 1931. It was more likely that the power to control the race's biological future would be misused either deliberately by the ruling class or short-sightedly by materialist ideologues. Birkenhead, whom Haldane accused of plagiarism, also predicted that the technology would be exploited by governments: 'If it were possible to breed a race of strong healthy creatures, swift and ductile in intricate drudgery, yet lacking ambition, what ruling class would resist the temptation?' In the end, though, it was *Brave New World* that would come to symbolize the distrust that Haldane's vision so often inspired. Here, Huxley created the model for an artificially designed society that ensured the superficial happiness of all by tailoring the individuals of each class to their work and ensuring that they were satisfied doing it. In some respects, the story went beyond Haldane – the 'Bokanovsky process' employed at the Central London Hatchery artificially divides fertilized eggs to produce clutches of identical humans who then have their development chemically retarded to provide just enough intelligence for menial work. They are then conditioned by psychological harassment to love the conditions they must live in. Sex is freely available because it has nothing to do with reproduction, and everyone is kept passive by a drug called 'soma'.[13]

Brave New World was immediately recognized as a powerful warning against the unrestrained use of the new biology. By the end of the 1930s, however, concern focused not on the mass production of inferior types, but on the eugenics movement's efforts to eliminate them. Those who favoured artificial intervention in human development saw things very differently. From Hollywood to Nazi Germany, there was a focus on efforts to develop a perfect human type, whether by eugenic selection or manipulation of development. As historian Christina Cogdell notes, in parallel with the modernist designers of physical objects they sought to 'streamline' the human race. MacFie – an opponent of eugenics – appealed to biochemical techniques to bring out our best qualities. Langdon-Davies thought the critics of *Brave New World* were unnecessarily frightened of the prospects for the scientific management of the race. He predicted that there would initially be a drive to ensure conformity, but eventually the state would see the need to encourage individuality.

The real enthusiasts went much further, arguing that to meet future challenges the human race would have to be completely redesigned. In his 1930 story 'The Last Judgement', Haldane predicted a future in which humanity would transform itself to achieve a lifespan of 3,000 years and ensure perfect happiness (at the expense of stagnation). Bernal's *The World, the Flesh and the Devil* predicted major developments in brain size and complete transformations of the body, including the creation of a new race adapted to space travel. Bernal explicitly downplayed the hope of genetic engineering, arguing that the control of individual development offered a more rapid way forward. The *Daily Herald*'s review of his book was titled 'Men with Ears under Lungs: What Humans may be like Another Day'. Stapledon's *Last and First Men* imagined a sequence of transformations, including the creation of the totally artificial 'fourth men' who pave the way for new forms adapted to life on other planets.[14]

Such ideas would soon become a staple of mainstream science fiction. But similar predictions continued to circulate more generally and became more acute in the 1960s as actual developments in biology began to accelerate. The eminent French biologist Jean Rostand's *Can Man Be Modified?* of 1959 quoted *Brave New World* and argued that the production of 'test tube babies' would allow somatic modifications to be used in addition to genetics. Haldane himself returned to the theme in a chapter for the 1963 CIBA Foundation volume *Man and His Future*, although by this time he had begun to focus on genetic selection. Gregory Pineaus, inspired by Haldane, worked toward the first oral contraceptive. For the pessimists, however, Gordon Rattray Taylor's *The Biological Time Bomb* of 1969 brought out all the worrying social and moral issues raised by these developments. He predicted a host of new biological technologies by the end of the century, including personality reconstruction, enhancement of intelligence, cloning and genetic engineering. What had begun as wild speculation now became an immediately threatening reality.[15]

All in the Mind

In Huxley's nightmare future, the lower ranks of society were psychologically conditioned to accept their lot by brutal applications of the behaviourist techniques developed by Watson. Behaviourism sought to free psychology from any notion of the mind as a separate entity from the body, thereby extending the materialist philosophy so distrusted by moralists. The same philosophy underpinned increasingly confident claims that a person's character and abilities could be manipulated by physical transformations of the brain brought about by surgery or the latest biochemical discoveries. Although these techniques were distrusted by conservatives, their proponents saw them as vehicles of social progress, helping to eliminating insanity and harmful personality traits.

We also need to recognize that this was not an age of uncontested materialism. Opposition to this philosophy was still active in many areas of science, most obviously in the new physics and in efforts to retain a vitalist or organicist perspective in the life sciences. Some biologists still hoped to show that life and mind were active components of nature, and this approach helped to sustain alternatives to materialism in the search for medical applications. Recognition of a psychological element in the campaign to defeat cancer has already been noted, an approach generalized in *Pygmalion*, R. M. Wilson's contribution on medicine to the 'Today and Tomorrow' series. He argued that stress was a major cause of most illnesses and that relief of stress would become the chief weapon in the armoury of the doctor of the future. This was why psychoanalysis and other unconventional therapies produced beneficial effects. The importance of the mind was also stressed by Alexis Carrell in his *Man the Unknown*, even though he had made his reputation in experimental medicine. But Carell went even further, expressing his conviction that paranormal phenomena such as clairvoyance and telepathy were real aspects of the active human mind. From a very different direction, the eminent physicist Sir Oliver Lodge became a leading proponent of spiritualism. He argued that the spirits of the deceased continued to exist on an ethereal plane, exploiting the world view (now increasingly discredited by relativity theory) which held that radio and other electromagnetic waves were transmitted by a tenuous material known as the ether. The 'Today and Tomorrow' series included a contribution by E. N. Bennett predicting that psychology would follow the lead shown by Lodge and expand to study paranormal phenomena, a view echoed in Gibbs's *The Day after Tomorrow*.

The fact that the brain and nervous system used electrical impulses allowed even those with materialist inclinations to take the possibility of telepathy seriously. Low argued that the brain might be able to broadcast like a radio and predicted that telepathy would become a fully developed human faculty in the future. His science-fiction novel *Mars Breaks Through* imagined interplanetary contact via artificially boosted telepathy rather than space travel. Julian Huxley was one among several eminent scientists to develop the theme – his own story 'The Tissue Culture King' describes how a renegade endocrinologist boosts the telepathic powers of an African tribe. In the 1930s, the psychologist William MacDougall, a prominent opponent of materialism, brought J. B. Rhine to Duke University to begin experiments designed to prove the reality of telepathy. Rhine's work was still going on in the post-war years, although increasingly marginalized from scientific psychology.[16]

The commentators who feared the intrusion of science into the study of human nature were, of course, concerned by the truly materialistic ideology in which the mind is merely a by-product of the brain's physical activity. Yet those who followed this path were convinced that it offered a path toward a better world in which human faculties would be augmented and mental illness banished. Modification of the brain either by chemistry or surgery could boost its activity or curb harmful behaviour patterns. The potential dangers were already apparent in Wells's *Island of Dr. Moreau*, but Moreau's efforts to boost the mentality of animals up to human levels were a forerunner of the more beneficial intrusions envisaged by the next generation of endocrinologists and surgeons.

In their collaboration, *The Science of Life*, Wells and Julian Huxley raised the prospect that chemicals might boost mental powers. Novelists lapped up the theme of brain enhancement. Noell Roger's *The New Adam* of 1926 and

H. L. Samuel's *An Unknown Land* of 1942 are two examples, the latter imagining surgical expansion of the skull to produce a higher type of humankind, a technique also suggested in a short story by S. Fowler Wright. For Low, the possibility of biochemical or surgical intervention offered the hope of curing insanity. This proposal was put into practice by surgeons such as Egas Moniz, who pioneered the technique of prefrontal lobotomy in 1935. Huge numbers of lobotomies were performed to 'correct' behavioural problems before the negative consequences became apparent in the following decade. Even then, Waldemar Kaempffert maintained the hope that more sophisticated surgical procedures might be successful, a prospect still recognized by Rostand in 1959. By this time, the emphasis was beginning to switch to the genes as the true foundation of the brain's structure and activity, but the prospect of personality reconstruction was still in play. Taylor's *Biological Time Bomb* predicted this, along with enhancement of intelligence and memory control by the end of the century.[17]

Physical control of the brain was not the only approach available to modify behaviour. Psychologists such as Watson abandoned the notion of the mind and treated the organism as a learning machine which could be programmed to behave in certain ways by appropriate conditioning. In *Brave New World*, the techniques used to create slaves adapted to their position in society were based on those used by Watson to condition the rats learning to run mazes in his laboratory. Yet we have seen that Watson himself envisioned wonderful schools where children raised apart from the family would be trained to become better human beings. Another behaviourist psychologist, B. F. Skinner, predicted the creation of a better society through well thought-out conditioning in his novel *Walden Two* of 1948.

The darker side was all too apparent, however, as illustrated the same year in Orwell's *1984*, where Winston Smith is conditioned to love Big Brother by sheer torture. The same prospect was explored in Len Deighton's *The Ipcress File* of 1962, made into a classic movie three years later. IPCRESS was the acronym of Induction of Psychoneuroses by Conditioned Reflex under strESS. Equal concern centred on the use of psychological techniques by governments and big business. Advertising was an obvious area where behaviour could be manipulated, and Watson himself worked on Madison Avenue after being forced from academia by a scandal. Governments could also seek to control their people through scientifically designed propaganda. By the time he published *Brave New World Revisited* in 1959, Huxley included advertising and propaganda among the misuses of psychology that he still saw as a danger to society. He thought the future would look more like that of his own prediction than Orwell's, but conceded that he had not foreseen the failure to control the expansion of the population and the uncontrolled multiplication of inferior types.

Breeding

Brave New World had a controlled population. But even in the 1930s, Huxley, like many intellectuals of the time, had been attracted to the arguments of the eugenics movement. Founded by Darwin's cousin, Francis Galton, this stressed the role of heredity as a determinant of character and warned against the uncontrolled breeding of the 'unfit'. Eugenics urged deliberate efforts to control human reproduction by boosting the number of children raised by the most able individuals and restricting the breeding of those with lower levels of intelligence. There is an extensive literature on the history of the movement, most of which focuses on the efforts to restrict the breeding of the social classes and races thought to harbour the least valuable characteristics. Some governments (including those of many states in the United States) took active steps by enacting legislation to sterilize the feebleminded and the insane. All too often, certain racial groups were targeted because it was assumed that they were genetically inferior. This was how it started in Germany too, but the Nazis eventually took the programme to its ultimate horrific conclusion by simply eliminating those deemed undesirable.[18]

The programmes of 'negative' eugenics were not necessarily based on any desire to create a new and superior form of humanity. They were designed to stem a tide of reproduction which, it was held, threatened to overwhelm the race and lead to mass degeneration. The plausibility of such fears did not depend on new scientific discoveries – the widespread assumption that heredity determines character pre-dated the emergence of genetics. And although some eugenists pointed to the relaxation of Darwinian natural selection in modern societies, the model they used was based on an analogy with the technique of artificial selection used by animal breeders long before Darwin borrowed it as a model for his theory. The rise of modern genetics was certainly used to provide evidence for the claim of hereditary determinism, but genetics itself was created on the basis of a deliberate exclusion of any role for the environment in shaping how the genes are manifested in the developing organism.

This hereditarian position was not universally accepted and we have seen how some of those who hoped to 'improve' individuals with hormones and other treatments expected the results to be transmitted to future generations. For reform eugenists such as C. W. Saleeby, it was crucial to improve the environment in which all were raised as well as seeking to restrict the breeding of the irredeemably unfit. Advocates of the inheritance of acquired characteristics (Lamarckism), including Paul Kammerer, proclaimed that improvements to the individual would become embedded in the race. The geneticists argued that modifications of individual development cannot be imposed on the genes, but there were many with liberal views which led them to insist that restricting the breeding of the superficially unfit would have little effect. For both Haldane and Julian Huxley, inequalities in society meant that poor conditions and education made it impossible to determine who might be genuinely carrying harmful genes. Eugenics would work in theory, but not in practice as long as poor characters might be the product of environment rather than heredity.

Those who held that heredity was the main factor were limited when it came to devising a programme of 'positive' eugenics to improve the race. They could only work within the limits provided by the best existing genetic characters. These would have to be concentrated by ensuring that the best-endowed individuals mated together instead of dissipating their advantages in the rest of the population. It was difficult to get the professional classes (who assumed that they were the reservoir of the fittest characters) to increase their family size. But at least they could be encouraged to marry only within the charmed circle, perhaps with some guidance through genetic testing of prospective spouses. Birkenhead saw this as the way ahead: 'This is the kind of eugenics which will develop in the future, rather than the obscurities of the human stud-farm which so many earnest, and generally unmarried, enthusiasts at present predict.' Carrell also preferred enlightened choice to compulsion – although habitual criminals 'should be humanely and economically disposed of in small euthanasic institutions supplied with proper gases'. Here we see how easily the eugenic movement could slide toward Fascism.[19]

Some thought that the state would take a hand at the top end of the scale by encouraging or requiring people to mate with genetically approved partners. As early as 1918, novelist Rose Macaulay depicted a world governed by a 'Ministry of Brains' which tested everyone to see who was fit to reproduce. Uncertified babies disappeared in mysterious circumstances. S. Fowler Wright also set a story in a future society governed by repressive eugenic policies. In his *Daedalus*, Haldane jokingly predicted: 'The eugenic official, a compound, it would appear, of the policeman, the priest and the procurer, is to hale us all off at suitable intervals to the local temple of Venus Generatrix with a partner chosen, one gathers, by something of the nature of a glorified medical board.' The process could, of course, be simplified if all births were to be ectogenic. Wells and Julian Huxley argued that although state control was not feasible at present, it would eventually be implemented.

One step further would be for the state to ensure that only a few of the most highly gifted males would donate the sperm used to fertilize the female population. Lockhart-Mummery predicted this would be the way chosen to improve the race, a view also expounded by geneticist H. J. Muller in his *Out of the Night* of 1936. Muller had gained his reputation by discovering that X-rays increase the rate of genetic mutation. Some hoped that this would lead to a process allowing control of how new genetic characters could be produced for exploitation by the eugenics programme. Bertrand Russell noted this possibility, even though he viewed the whole idea of controlled reproduction with alarm. Eventually, the prospect of breeding new forms of humanity would appear in mainstream science fiction, including Robert Heinlein's *Beyond This Horizon*, originally published in 1942. The possibility of genetic engineering was thus envisaged long before geneticists had any real understanding of the nature of the gene.[20]

Following the lead offered by Bernal, science fiction writers could imagine genetically engineered races produced for new environments such as space travel. Here on Earth, there was no clear consensus as to how the future human race would be shaped. Many passively assumed that the ideals of their own race and social class would be realized, perhaps in a 'streamlined' form. A eugenic tract by Charles Wicksteed Armstong had a chapter on 'The Destiny of Man' based on Lodge's vision of a more spiritual future. Such expectations could easily be consolidated into something more clearly defined, leading to news-paper headlines such as one in the *New York Times* in December 1929: 'Science Pictures a Superman of Tomorrow'. The Nazis' idealization of the blond, blue-eyed Aryan is the most notorious example of this trend, to be achieved both by manipulating the breeding of the favoured race and supressing that of alien types.

Some environmentalists hoped to breed a race that would be adapted to a more natural way of life. A few eugenists recognized that trying to define an artificial goal was pointless in an ever-changing world. Schiller's *Tantalus* argued that since there was no law of progress in evolution, the human race would have to preserve a range of variation even in an upgraded state. Another contribution to the 'Today and Tomorrow' series by the leading British eugenist C. P. Blacker anticipated a point made in *Brave New World* – someone has to do the menial work, so there was no point breeding a race composed solely of the intellectually gifted.[21]

None of the fanciful schemes for breeding supermen came to fruition, but the harsher aspects of negative eugenics became only too apparent in the 1930s. Many states in the United States enacted legislation for the compulsory ster-ilization of the feebleminded, and Nazi exhibitions on racial hygiene toured the country freely. In the end, the Nazis simply began to kill off those who they wished to exclude from the future of the race. American eugenics also devel-oped a strong focus on race, fearing the unrestricted breeding of 'unfit' immi-grants. When the horrors of the Nazi death camps were revealed to the world in 1945, explicit talk of eugenics and race science fell silent. But the underlying prejudices remained active under the surface, concealed in the form of growing concerns about the explosive growth in the world's population. This was not a new issue, as we saw in the previous chapter. While some worried about the decline in the population of European countries (and Langdon-Davies even predicted a decline worldwide), many were concerned that the benefits of scientific agriculture and medicine would be offset by a rise in the population of the underdeveloped world.[22] Already in the interwar years, many feared that resources would be stretched as the non-white races outbred the populations of the developed countries.

Such concerns fell into the background in the 1940s, but soon resurfaced in the age of the Cold War as the West and the Soviets sought to extend their influence. As the rival power blocs sought to convince the third world that their systems offered the best hope of progress, many in the West began to articulate the old Malthusian concern that the benefits of scientific progress would be gobbled up by an expanding population. William Vogt's *Road to Survival* of 1948 highlighted the damage that was being done to the environment by the over-exploitation of resources, but also insisted that the demands were often driven by the expanding population. He suggested that aid to the developing world should be made contingent on the governments adopting policies to promote voluntary contraception.[23]

Another early expression of these fears came in Sir Charles Galton Darwin's *The Next Million Years* of 1952, reiterated in his 1958 Rede Lecture *The Problems of World Population*. Darwin was well aware of the potential developments in energy production, and he knew better than to repeat the old arguments for eugenics and race science (although his true feelings emerged when he imagined a future African historian commenting on the decline of the white race). He also recognized that a nuclear war could destroy all hope of progress. His real concern was that without some system to control reproduction, any progress in food production would simply lead to further population expansion and the ultimate exhaustion of the planet's natural resources. His views were echoed in Harrison Brown's *The Challenge of Man's Future* in 1954. Both were pessimistic about the prospects for changing the world's attitudes toward reproduction.[24]

Aldous Huxley raised the problem of population in his *Brave New World Revisited*, arguing that medical advances only seemed to make things worse. In the 1960s, these concerns began to hit the headlines as they were highlighted in books such as Georg Borgstrom's *The Hungry Planet* and Paul Ehrlich's *The Population Bomb*. Borgstrom stressed that all efforts to use technology to tackle the problem of poverty had failed. A new direction was needed to integrate research on global issues. We certainly shouldn't be wasting money developing foods for astronauts when most of the world was starving. Ehrlich echoed these views, arguing that efforts to expand the food supply were resulting in the destruction of the planet. Environmentalism was a growing concern (highlighted, paradoxically, by Apollo 8's photographs of the Earth from space). Further developments in agricultural science were leading to the 'Green Revolution'. President Lyndon Johnson's 1965 State of the Union address dedicated his administration to solving the problems of population and resources, highlighted three years later in the Club of Rome's first predictions. In the end, however, none of the doomsayers offered policies that might replace the earlier calls for governments to impose restrictions on the right to have children.

We thus end with an apparent paradox: just as writers such as Jean Rostand and Gordon Rattray Taylor were proclaiming (and warning about) the wonders that would be performed by the biological sciences acting on individuals, another set of writers was alerting the world to the danger that all of the scientists' efforts would be negated by the apparent impossibility of stopping people from having children. The discovery of the structure of DNA by Watson and Crick in 1953 seemed to open up the prospect that now, at last, we might be able to control the genes at a level only dreamt of by the eugenists of the previous generation. In 1966, a book edited by Lawrence Lessing appeared with the title *DNA: At the Core of Life Itself*. It would be some considerable time before the workings of the gene would be understood at a level that would permit real control, but the potential was now being recognized. Still, the tension between the urge to control individual human life and the concern that no rational plan is available to spread the potential benefits beyond the developed world remains to haunt us today.[25]

12   Epilogue
Plus ça change?

Some things certainly did change in the half century or so separating the periods in which H. G. Wells and Isaac Asimov made their names imagining future worlds. Science fiction emerged as a distinct and increasingly popular component of modern culture. Other ways in which information and comment on science and technology were presented to the public underwent a similar transformation. At the turn of the century, popular science writing focused mainly on educating the public about established achievements. During the interwar years, it was increasingly aimed at providing news of the latest developments and their potential implications. Some authors emphasized the benefits, others worried about the hazards, but all were convinced that everyone's lives would be transformed by what was becoming available. Inevitably, they sought to imagine what the next steps would be. By the 1960s, writers such as Asimov and Arthur C. Clarke had gained worldwide reputations for their forecasts – Asimov's predictions for the 2014 World's Fair featured in the *New York Times*. Futurology now became more organized as groups such as the RAND Corporation sought to influence global ideological conflicts. The World Future Society began publishing its *Futurist* magazine in 1967 and books such as Dennis Gabor's *Inventing the Future* became bestsellers. In 1970, Alvin Toffler's *Future Shock* drove home the need to take futurology seriously.[1]

The expansion in the level of public expectation was driven by the increasingly obvious way in which new technical developments had transformed society. As Wells was making his name, the discovery of new phenomena such as X-rays, radio waves and cathode rays were revolutionizing physics and opening up a world of new technologies. Biologists were transforming our understanding of how the body works and how heredity controls individual characters. The cinema, electric lighting and motor cars were already providing new experiences to the public, soon to be joined by radio broadcasting and heavier-than-air flight. By the 1960s, the all-electric home had become commonplace, radio was being challenged by television, and rapid air-transport was widely available. Space travel and atomic power – once widely dismissed as moonshine – were now a reality. At the same time, the atomic bomb and intercontinental missile provided ammunition for the pessimists who claimed that the benefits of technological advance were far outweighed by the creation of new and potentially catastrophic hazards.

The processes by which these innovations were conceived, invented and (when successful) made available are the prime area of interest for historians of science and technology. In some cases, new possibilities seemed to present themselves as more-or-less obvious extensions of existing techniques. A public already familiar with cinema and radio broadcasting expected television as the next step and a host of inventors strove to produce it for them. The same was true for mass air-transport, but here too there were rival systems on offer and the conflict between the supporters of airships, land-based aeroplanes and flying boats offers a classic illustration of the uncertainties involved in turning speculation into reality. The hopes for supersonic flight for all and personal aircraft shows how over-enthusiastic expectation all too easily pushes the boundary beyond what makes sense in the commercial world. Some short-lived fads such as death rays, radium cures for all ills and rejuvenation were generated by publicity encouraging hopes far beyond what was plausible given the science of the day. When innovations did work out, all too often the driving force was military rather than civilian interest, and a few key inventions including radar and the jet engine emerged from this background outside the range of popular expectation. And then there is the huge list of inventions based on sound science and engineering which didn't catch on in the real world.

There was immense technological progress, however messy the transition from invention to introduction may have been. But this book has addressed a different set of concerns focused more on the ways in which speculation about the future throws light on public perception of science and technology. Here, there is much more continuity than we might imagine. When we look at what people wanted and expected from science, what the innovators thought would be likely new developments, and what conservative thinkers feared as the downsides of progress, we find that many of the features we associate with the late twentieth century and the modern world were anticipated in earlier decades. There were changes of course: the 1960s saw an outburst of concern that technology was threatening our very existence, a concern that was more intense than anything that had gone before. But distrust of the effects of science and technology had begun in the Industrial Revolution and was extremely active in the interwar years. The optimists also became more organized in their efforts to predict the future in the post-war era, but again their activities were merely an intensification of what had gone before. And – allowing for those predictions that had actually been fulfilled – their expectations in the 1960s were still surprisingly similar to those articulated in previous decades.

The organized futurology of the 1960s can be seen as the intensification of an activity that was operating throughout the first half of the century. This earlier wave of prediction went far beyond the doom-saying of the novelists who imagined future wars or nightmare societies enslaved by technology. This book has surveyed a vast range of literature produced by scientists, engineers and popular writers who were already taking the business of prediction seriously. From major scientists such as Haldane and Bernal, through self-styled experts such as A. M. Low, to a plethora of writers with varying degrees of technical knowledge, there was a constant stream of books and articles speculating on the next wave of developments and their potential implications. The huge number of books published during the 1920s in the 'Today and Tomorrow' series provided one very visible focus for this activity. Titles of other studies such as Birkenhead's *The World in 2030 A.D.* and Langdon-Davies's *A Short History of the Future* are typical examples of individual contributions to the genre. The situation changed in the 1960s because governments and industrial concerns began to take the future seriously, employing those willing to take the risk of prediction to do things they had already been doing for decades on an individual basis.

A few of the predictions made in the 1960s were new, in some cases exposing blind spots in what had gone before. By this time, computers had appeared and were raising the prospects of less drudgery in the office and fears of a more controlled society. The invention of the transistor allowed the miniaturization of all electronic gadgets and raised the possibility of small computers in the home. The latter advance was anticipated by Herman Kahn and Anthony Wiener in 1967, although most experts still preferred the older idea that we would all have home terminals connected to a single giant computer. The more open-ended world of the internet was a conception that would gradually emerge over the next few decades. A surprisingly large proportion of the predictions made at this time were, however, merely the residue of those made in previous decades that had not yet been realized. Whether we look at Kahn and Wiener's predictions for the year 2000, Nigel Calder's 1964 collection of articles anticipating the real world of 1984, or Asimov's list of what he might expect at the World's Fair of 2014, we see many of the items already on the wish lists of earlier writers. Plentiful atomic power, perhaps even small-scale atomic power units, personal flying machines, colonies on the moon and on Mars, synthetic food, techniques for extending life, these and many other predictions reappear over and over again. Far from representing a failure of the imagination, this repetition confirms that the same kind of people – in some cases the same individuals – were simply continuing the activity of previous decades.

There were some bolder prophets: in his *Profiles of the Future* of 1962, Arthur C. Clarke imagined a space drive that worked by modifying gravity (although that idea had also featured occasionally in earlier predictions). He also thought there might be instantaneous travel and the production of material by replication of atomic structures, ideas that would reappear in *Star Trek* (and still seem out of sight today). The visionary designer Gerald K. O'Neill would soon publish details for projected space colonies. These hopes were inspired by the unexpected success of the space race, but were soon dashed by the rapid loss of public interest in the Apollo programme and manned space-flight in general.[2]

For the enthusiasts, the post-war era seemed to open up the path toward a new world of science-driven innovation that offered unlimited benefits to mankind. They were not unaware of the dangers represented by the threat of nuclear war and environmental pollution, but – provided all-out war could be avoided – they still imagined that more technology would solve the world's problems. This attitude was manifest in C. P. Snow's claim in *The Two Cultures* that the scientists had the future in their bones.

The rise of an active movement critical of science and determined to block or reverse many of its applications came as something of a shock. It seemed a new development, at least in its intensity. In November 1962, *New Scientist* magazine published an editorial entitled 'Science in Disrepute', lamenting the rising tide of public disappointment and anger. In America, the botanist Barry Commoner, who had been deeply involved with the growing recognition of the dangers of nuclear fallout, published his *Science and Survival* to address the fear that the process of technological development was getting out of hand. Ralph Lapp's *The New Priesthood* compared science to a religion and lamented 'The Tyranny of Technology'. There was a growing feeling that the scientific community had lost its integrity by selling out to governments and big business. Commoner also suggested that although the dangerous by-products of industrial development had always been apparent, new threats such as fallout and environmental pollution were invisible and hence seemed all the more threatening.[3]

The population boom presented another problem, driven by the very improvements in medical science that had extended the life expectancy even of the masses in the third world. This too seemed new: in 1963, Dennis Gabor's *Inventing the Future* noted the unexpectedness with which the issue of population exploded onto the scene after the war. Gabor conceded that writers such as A. M. Carr-Saunders had raised the issue of unchecked population expansion between the wars, but dismissed their concerns as irrelevant because they had thought that the population of the developed world was falling. This was true, but hardly justified the claim that the big issue had not been recognized earlier since it was still the growing population of the underdeveloped world that was the main worry. Karl Sax's *Standing Room Only* of 1955 used the same title as Ross's book of 1928, lacking only the question mark.[4]

In this and some other cases, it is possible that the hiatus of World War II temporarily deflected attention away from the problem. The same issues re-emerged into the public consciousness after the war, but were reinforced by the growing concerns about the growth of what became known as the military-industrial complex. Science and technology had been increasingly involved with government and industry before the war, but the scale of the links had now expanded dramatically. As the public became more worried about the threats of nuclear war, overpopulation and pollution, the deliberate involvement of science and technology in the creation of the threats was highlighted. In some respects, the situation was indeed new, but the threats themselves had often been identified decades earlier when the reputation of science had seemed less tarnished.

Our survey has uncovered earlier manifestations of the fears that resurfaced in the 1960s. Even the atomic bomb and its devastating consequences had been predicted by Wells and several later fiction writers. The prospect that civilization and perhaps even the human race might be destroyed in some future conflict was commonplace in novels through the interwar years. The combination of aviation, poison gas and germ warfare was seen as having the potential to make whole areas and perhaps the whole planet uninhabitable. The atomic and hydrogen bombs made the prospect seem more realistic, but as Harold Macmillan noted, the pre-war fears seemed genuine enough at the time. Fallout and the new chemical insecticides added to the threats faced by the environment and, as Commoner suggested, their insidious nature made them seem all the more worrying. But as he also pointed out, the dangerous by-products of industry had been obvious for centuries. The Dust Bowl confirmed that the effects of industrialized agriculture could be devastating. Concerns that wild species were being driven to extinction by human activity also become more widespread in the early twentieth century, along with calls for environmental protection. The 1960s saw expectations rise that DNA would unlock the secrets of life, with potentially both exciting and dangerous consequences. But those consequences had already been pointed out by Haldane and others, and were widely debated from the 1920s onwards. The renewed concerns were driven by the expectation that the science to make these effects possible was now closer to hand.

Historians may disagree over just how new the situation in the 1960s was. But one element of continuity seems incontrovertible: there was an ongoing debate between those who welcomed the benefits offered by technological change (whatever the dangerous by-products) and conservatives who were indifferent to the benefits and feared the destruction of traditional values and ways of life. In his *Technopolis* of 1969, Nigel Calder addressed the popular fears and distinguished between the attitudes of the 'mugs and zealots'. A mug (short for mugwump, which Calder took to mean a stick-in-the mud) was a tender-minded conservative ever harping on about the potential dangers. The zealots were those whose enthusiasm for innovation overrode all other concerns, including the ever-rising tide of harmful consequences. Like Commoner, Calder urged better control of the applications of science, with the scientists themselves playing a more socially conscious role.[5]

Our study of the early twentieth century has shown that the conflict between the mugs and the zealots long pre-dates the 1960s. The balance between the two has shifted across time and place, with one side seeming more active at one point only to see the situation change with further technical and social developments. But both sides are always present, and it is a mistake to focus on one or the other exclusively if we wish to gain a balanced picture of what was going on. The source of evidence is crucial. Those such as I. F. Clarke who approach the field of prognostication primarily through the writings of literary figures gain the impression that it has been dominated throughout by the pessimists. Acknowledging the input from popular science writing and pulp science fiction allows a much more nuanced picture to emerge.

This point applies equally well to efforts to identify a supposed consensus for a particular period or country. Fear of new weapons and their devastating consequences fuels the argument that the interwar years were a 'morbid age' obsessed with the threat of collapsing civilization. But the future-war novels of the 1930s were primarily written by Europeans fearful of a war that everyone knew was coming, and even there the popular science literature was full of expectations of future benefits. In America, those benefits were being celebrated at the New York World's Fair even as the war the Europeans had feared actually broke out. Optimism when the war was eventually won produced a temporary boost in the hopes for science-driven progress, hopes that were only too soon dashed by the rising tide of harmful applications and by-products.

Attitudes to science reflect changing circumstances, but they also reflect professional interests and ideological commitments. Literary figures and moralists tend to be suspicious, while the scientists and engineers actually engaged in the process of development tend to be enthusiastic, perhaps uncritically so. Attitudes may also be reflected in the age of the writer, a point I am acutely conscious of, having been a technophile for much of my life, but now increasingly resistant to the expanding world of the internet and social media. I have moved from being something of a zealot to the position of a confirmed mug. I hope that seeing both sides of the question has helped me to achieve a balance in the study presented here.

Notes

1 Introduction

1. Isaac Asimov, 'Visit to the World's Fair of 2014', *New York Times*, 16 August 1964, available at http://www.nytimes.com/books/97/03/23/lifetimes/asi-v-fair.html (accessed 22 February 2016).
2. Benford (ed.), *The Wonderful Future that Never Was*. An earlier and equally well-illustrated survey is Corn and Horrigan (eds), *Yesterday's Tomorrows*. For a more light-hearted questioning of why promised technologies have failed to materialize, see Wilson, *Where's My Jetpack?*
3. Benford *et al.* (eds), *The Amazing Weapons that Never Were*.
4. Clarke, *The Pattern of Expectation*, chs 8 and 9; and the same author's *Voices Prophesying War*.
5. Overy, *The Morbid Age: Britain and the Crisis of Civilization*.
6. Orwell, 'Inside the Whale' [1940] in *The Collected Essays, Journalism and Letters of George Orwell*, I: 493–527, p. 507 (emphasis in the original).
7. Blom, *The Vertigo Years: Change and Culture in the West, 1900–1914*; and Panchasi, *Future Tense*.
8. On America in the 1930s, see Bush, *The Streamlined Decade*. These national differences are discussed in more detail below.
9. Detailed references will be given when these authors are introduced in Chapter 2. In most cases, the studies focus on the literary quality of the works, although the sources of information about science and technology are sometimes considered. On Wells's relationship to these figures, see Hillegas, *The Future as Nightmare*.
10. Aldridge, *The Scientific World View in Dystopia*, Appendix. See also Chapter 2 below.
11. For a survey by a real science fiction author, see Aldiss and Wingrove, *Billion Year Spree*. For scholarly studies, see Carter, *The Creation of Tomorrow*; Cheng, *Astounding Wonder*; Stover, *Science Fiction from Wells to Heinlein*; and Haynes, *From Faust to Strangelove*. General works include James and Mendlesohn (eds), *The Cambridge Companion to Science Fiction*; Link and Canavan (eds), *The Cambridge Companion to American Science Fiction*; Nichols (ed.), *The Encyclopedia of Science Fiction*; Westfahl, *The Mechanics of Wonder*; and the same author's *Cosmic Engineers*.
12. Frayling, *Mad, Bad and Dangerous?* See also Brosnan, *Future Tense*; and Carpenter, *Dramatists and the Bomb*.

13. Asimov, 'A Literature of Ideas' in his *Towards Tomorrow*, pp. 161–9, see p. 167. See also *Asimov on Science Fiction*, pp. 18 and 82.
14. Clarke, *Profiles of the Future*, p. 10 (emphasis in the original).
15. See Carter, *The Creation of Tomorrow*.
16. On American popular science, see La Follette, *Making Science Our Own*; and on Britain, Broks, *Media Science before the Great War* and my own *Science for All*. More generally, see Broks, *Understanding Popular Science*; and Whitworth, *Einstein's Wake*. On science broadcasting, see La Follette, *Science on the Air*; and on documentary films, see Boon, *Films of Fact*.
17. Wells's *Anticipations* of 1901 is pure futurology. In addition to Asimov's 1964 predictions, see also his *Living in the Future*; and for Clarke, see his *Exploration of Space* and *Profiles of the Future*.
18. For instance, John Hodgson's 1929 book *The Time-Journey of Dr Barton*, which depicts life in the year 3927. The same technique was, of course, used by Wells, but Hodgson's book explicitly claims to be (in the subtitle) *An Engineering and Sociological Forecast based on Present Possibilities*. Chapter 11 of Lockhart-Mummery's *After Us* of 1936 is a fictional account of a New Zealand family's visit to London in 2456 AD.
19. Wells, *The Discovery of the Future*, pp. 8–10 and 24 (originally a lecture delivered at the Royal Institution). See also his *Experiment in Autobiography*, vol. 2, pp. 648–9.
20. This lecture was later reprinted in 1969 in *The Two Cultures and a Second Look*.
21. See Snow, *The Two Cultures and a Second Look*, which offers also some later reflections on the debate. See Ortolano, *The Two Cultures Controversy*; and on Snow's distorted image of government spending, Edgerton, *Warfare State*. The school, by the way, was Alderman Newton's Boys School, Leicester.
22. Peter Broks's *Media Science before the Great War* notes how living scientists were usually depicted sympathetically, while fictional scientists were always deranged loners out to conquer or destroy the world. See also Haynes, *From Faust to Strangelove*; and Clarke, *Voices Prophesying War*.
23. Snow, *Science and Government*, pp. 68–9.
24. Clarke is also the author of *Voices Prophesying War*. The future war novels are discussed in more detail below (Chapters 2 and 9).
25. Hobsbawm, 'C (for Crisis)'.
26. In date order of the first publication, these are by: A. M. Low (1925 and 1934); Henry Ford (1926); Philip Gibbs (1928); Charles A. Beard (1928 and 1930); Stuart Chase (1929); Robert Millikan (1930); the Earl of Birkenhead (1930); Bertrand Russell (1931); I. O. Evans (1933); Ritchie Calder (1934a and 1934b); Jonathan N. Leonard (1935); C. C. Furnas (1936); J. Langdon-Davies (1936); J. P. Lockhart-Mummery (1936); Gerald Wendt (1939); and Waldemar Kaempffert (1940). Details are given in the bibliography.
27. Haldane's *Daedalus*, Bernal's *The World, the Flesh and the Devil* and Russell's *Icarus* appeared in this series, along with over twenty other books on scientific or technical topics, many of which are mentioned in the course of the following chapters.
28. See also the 1940 survey by Graves and Hodge, *The Long Week-End*.

29. See Akin, *Technocracy and the American Dream*; Ritschel, *The Politics of Planning*; and, on international planning, Mazower, *Governing the World*. I will return to this topic briefly in Chapters 4 and 9.
30. Examples of this literature include Branford, *Science and Sanctity*; Cram, *Towards the Great Peace*; Cranmer-Byng, *Tomorrow's Star*; Dreaper, *The Future of Civilisation and Social Science*; Ingram, *The Coming Civilization*; Marvin, *The New Vision of Man*; and Wates, *All for the Golden Age*. On the role of liberal Christianity, see Bowler, *Reconciling Science and Religion*. From France, see Le Bon, *The World in Revolt*; from Holland, Huizinga, *In the Shadow of Tomorrow*; and, from Germany, Rathenau, *In Days to Come*.
31. See, for instance, Joad (ed.), *Manifesto*, which outlines the plan of the Federation of Progressive Societies and Individuals, to which Wells and Stapledon (among others) contributed.
32. Freud, *The Future of an Illusion*.
33. On evolutionary biologists' continuing interest in the idea of progress, see Michael Ruse, *Monad to Man*; and Bowler, *Monkey Trials and Gorilla Sermons*. On popular representations of evolution in America, see Clark, *God – or Gorilla*.
34. On the influence of technology on American life, see Steven Cassedy, *Connected*; and on futurology, see Corn (ed.), *Imagining Tomorrow*; and Segal, *Technological Utopianism in American Culture*. More generally, see Currell, *American Culture in the 1920s*; and Eldridge, *American Culture in the 1930s*. For a snapshot of the situation in 1927, there is the popular survey by Bryson, *One Summer* and on the following years Bush, *The Streamlined Decade*. There are useful comments on the differences between British and American attitudes in Parrinder, *Shadows of the Future*, ch. 9. On France, see Panchasi, *Future Tense*.
35. See Gooday, *Domesticating Electricity*. Another good example is the uncertainty over the introduction of steamships in the mid nineteenth century, discussed in ch. 3 of Marsden and Smith, *Engineering Empires*.
36. This point is made, again in the context of electrical technology, by Arapostathis and Gooday, *Patently Contestable*.

2 The Prophets

1. See Segal, *Technological Utopianism in American Culture*, esp. ch. 3; and Carlson, *Tesla: Inventor of the Electrical Age*. For Astor's vision of the Earth in the year 2000, see book 1 of his *A Journey in Other Worlds*.
2. On Wells as a popular science writer, see Bowler, *Science for All*, especially pp. 103–7.
3. *An Experiment in Autobiography*, vol. 2, p. 645.
4. See the introduction to the Penguin edition of *The Sleeper Awakes*. There are many studies of Wells's futuristic romances, including Bergonzi, *The Early H. G. Wells*; Haynes, *H. G. Wells: Discoverer of the Future*; Huntington, *The Logic of Fantasy*; and Parrinder, *Shadows of the Future*. Haynes in particular stresses the role of Wells's interest in science, downplaying suggestions that this was largely overlaid by his romanticism.
5. *An Experiment in Autobiography*, vol. 2, p. 645.
6. See Wells's introduction to the 1914 reprint of *Anticipations*, p. ix.

7. *Anticipations*, pp. 301 and 317. Wells insisted, though, that the common prejudice against the Jews was unjustified.
8. On Brennan's monorail, see *The War in the Air*, pp. 12–14 and the image facing p. 14. It is described more fully in Chapter 6 below.
9. *The World Set Free*, p. 257.
10. 1932, p. 812.
11. *The Shape of Things to Come*, Book 4, 'The Modern State Militant'.
12. On the movie version, see Leon Stover, *The Prophetic Soul* and the text edited by Stover cited as Wells, *Things to Come*. For the concluding scenes, see the latter pp. 197–204.
13. *The Shape of Things to Come*, pp. 284–5.
14. In Kipling, *Actions and Reactions*, pp. 111–67, see p. 120 and for the ABC's motto, pp. 135–6.
15. The story is reprinted in Kipling, *A Diversity of Creatures*, pp. 1–44. For a description of how cities have become dispersed into vast collections of private estates, see the conclusion.
16. See Hillegas, *The Future as Nightmare*; Clarke, *The Pattern of Expectation*, chs 5–9; and Aldridge, *The Scientific World View in Dystopia*.
17. 'The Machine Stops' is pp. 109–46 of the edition of Forster's *Collected Short Stories* cited in the bibliography.
18. Stapledon, *Star Maker*, pp. 34–8. On Stapledon's writings, see the detailed studies by Robert Crossley and Leslie A. Fielder.
19. See Brosnan, *Future Tense*, pp. 29–30; and Baxter, *Science Fiction in the Cinema*, ch. 3.
20. Clarence Brown's introduction to the new translation of *We* cited in the bibliography gives useful information on the background to Zamyatin's thought. On Soviet attitudes to science, see Krementsov, *Revolutionary Experiments*.
21. These projections are discussed later in this chapter. Popular and fictional accounts are discussed in: Armstrong, *Modernism, Technology and the Body*; Krementsov, *Revolutionary Experiments*; McLaren, *Reproduction by Design*; and Sengoopta, *The Most Secret Quintessence of Life*.
22. On Huxley's career, see Murray, *Aldous Huxley*; ch. 21 is on *Brave New World*.
23. This is quite extensive – see pp. 150–72 of the edition of *1984* cited.
24. Aldridge, *The Scientific World View in Dystopia*, Appendix.
25. See the note prefaced to Lewis's *Out of the Silent Planet*. He was also opposed to Haldane and Stapledon's visions of the future; see Green and Hooper, *C. S. Lewis: A Biography*, p. 163.
26. Wells, *The World Set Free*, ch. 2.
27. Stapledon, *Last and First Men*, pp. 20 and 31–2.
28. Examples include William Le Queux's *The Invasion of 1910* and Erskine Childers's *The Riddle of the Sands*, both of which postulate German invasions of Britain by sea (thwarted in the latter case). On the future war novels, see Clarke, *Voices Prophesying War* and the same author's *The Pattern of Expectation*. On the interwar years, see Ceadel, 'Popular Fiction and the Next War, 1918–39'.
29. For details on the many American novels and Edison's involvement, see Franklin, *War Stars*, chs 2 and 3.
30. Norton, *The Vanishing Fleet*, pp. vii–viii.

31. Bywater, *The Great Pacific War*. It is thought that Japanese naval planners may have read the book and taken note of its warnings.
32. See Haapamaki, *The Coming of the Aerial War*; and Holman, *The Next War in the Air*. For more details on these novels, see Chapter 9 below.
33. Nicolson, *Public Faces*; and Priestley, *The Doomsday Men*.
34. Wyndham, *The Midwich Cuckoos*; and Miller, *A Canticle for Leibowitz*.
35. Aldiss and Wingrove, *Billion Year Spree*, pp. 251–2. For other studies of the topic, see Carter, *The Creation of Tomorrow*; Cheng, *Astounding Wonder*; James and Mendlesohn (eds), *The Cambridge Companion to Science Fiction*; Nicholls (ed.), *The Encyclopedia of Science Fiction*; and Stover, *Science Fiction from Wells to Heinlein*. On magazines specifically, see the introductions to Ashley (ed.), *The History of the Science Fiction Magazine*, parts 1 and 2.
36. For the Gernsback quotation, see Carter, *The Creation of Tomorrow*, p. 11. The Campbell quotation is from Aldiss and Wingrove, *Billion Year Spree*, pp. 274–5. Aldiss also quotes Arthur C. Clarke's point that '[a]ny sufficiently advanced technology is indistinguishable from magic', which shows how difficult it was to maintain Gernsback's principle, see p. 281. The science consultants for *Science Wonder Stories* are listed at Cheng, *Astounding Wonder*, p. 102. Most have academic qualifications and many occupied professional positions.
37. Pohl, *The Way the Future Was*, p. 180.
38. See Westfahl's introduction to Westfahl *et al.* (eds), *Science Fiction and the Prediction of the Future*. For details of these authors, see also Westfahl's *Cosmic Engineers* and *The Mechanics of Wonder* and Chen, *Astounding Wonder*.
39. Orwell, 'Boys' Weeklies' [1940] in his *Collected Essays, Journalism and Letters*, vol. 1, pp. 460–85, see p. 475; 'Frank Richards Responds to George Orwell', ibid., pp. 485–93, see p. 492. Richards is best known as the author of the 'Billy Bunter' stories. My own collection contains an obscure comic, the *Boys' Magazine*, which began a serial 'The War in Space' in 1926; see Fig. 8.2, p. 142.
40. On Dan Dare and the *Eagle*, see Morris (ed.), *The Best of Eagle*; Morris and Hallwood, *Living with Eagles*; and Watkins, 'Piloting the Nation: Dan Dare in the 1950s'. On the illustrator Frank Hampton, see Crompton, *The Man Who Drew Tomorrow*. More generally, see Chapman, *British Comics: A Cultural History*.
41. For Asimov's view that his fiction was an extension of reality, see his 'Escape from Reality' in his *Is Anyone There?*, ch. 32. See also his *The Martian Way and Other Stories*. For details of his career, see the revised version of his autobiography, *I. Asimov: A Memoir*. See also Fiedler and Mele, *Isaac Asimov*; and Olander and Greenberg (eds), *Isaac Asimov*. Asimov, Heinlein and Clarke became known as the 'big three' of science fiction. On Heinlein, see Franklin, *Robert A. Heinlein: America as Science Fiction*; and Olander and Greenberg (eds), *Robert A. Heinlein*. On Clarke, see his autobiography, *Astounding Days*; McAleer, *Odyssey: The Authorized Biography of Arthur C. Clarke*; and Olander and Greenberg (eds), *Arthur C. Clarke*.
42. On the subculture of the radio hams, for instance, Douglas, *Inventing American Broadcasting*, p. xxii and ch. 6.
43. Pohl, *The Way the Future Was*, chs 4 and 5.

44. Cheng, *Astounding Wonder*, ch. 8. Clarke's *The Exploration of Space* has a diagram of how geosynchronous satellites would transmit to the whole earth on the rear endpapers; see Fig. 5.2, p. 81.
45. On American popular science, see La Follette, *Making Science our Own*; and on Britain, see Broks, *Media Science before the Great War* and my own *Science for All*. For more details, see also my 'Popular Science Magazines in Interwar Britain' and 'Discovering Science from an Armchair'. On German magazines, see Schirmacher, 'From *Kosmos* to *Koralle*'. Radio broadcasting also became important; see La Follette, *Science on the Air*; and on documentary films Boon, *Films of Fact*.
46. 'Our Policy', *Practical Mechanics*, 1 (October 1933): 3.
47. I will provide details of the magazine's predictive contents in a forthcoming paper for *BJHS Themes*. See also Manduco, *The Meccano Magazine*.
48. Details of British scientists writing for a wider readership are given in Bowler, *Science for All* and 'Popular Science Magazines in Interwar Britain'. On American scientists' efforts, see La Follette, *Making Science our Own*; and Tobey, *The American Ideology of National Science*.
49. On Low, see Bloom, *He Lit the Lamp*. J. G. Crowther, who became science correspondent of the *Manchester Guardian*, also got his science training during the war.
50. On atomic energy, see the anonymous 'World's Most Terrible Secret', *Tit-Bits*, 3 July 1920, p. 362; and 'The Electron's Hidden Magic', 7 March 1925, p. 32. See also Gernsback, 'Fifty Years from Now', published 14 March 1925.
51. On atomic energy in the *Illustrated London News*, see Wall, 'Seeking to Disrupt the Atom: Immeasurable Energy'. Death rays feature in anonymous reports 26 April 1924, pp. 733–5, while stills from Lang's movie *The Woman in the Moon* appeared on 24 August 1924, pp. 346–7 and 2 November 1924, p. 701.
52. Reprinted in Churchill, *Thoughts and Adventures*, pp. 174–8 and 193–204; see Farmelo, *Churchill's Bomb*, pp. 41–3. The predictions were repeated for the *News of the World* in 1937, see Farmelo, pp. 87–8. Churchill also knew the Earl of Birkenhead, whose *The World in 2030 A.D.* is mentioned below. See also Inge, 'The Future of the Human Race'.
53. On world fairs, see Greenhalgh, *Ephemeral Vistas*; and Kargon et al., *World's Fairs on the Eve of War*. On the British Empire Exhibition, see Donald R. Knight and Sabey, *The Lion Roars at Wembley*; and Crampsey, *The Empire Exhibition of 1938*. On the Festival of Britain, see Banham and Hillier (eds), *A Tonic to the Nation*; and Forgan, 'Festivals of Science and the Two Cultures'.
54. Dronamraju's *Haldane's Daedalus Revisited* reprints the original text along with several commentaries. On Haldane's career, see Clark, *JBS*.
55. Adams, 'Last Judgement: The Visionary Biology of J. B. S. Haldane'. For the original story, see Haldane, *Possible Worlds*, pp. 278–312.
56. Haldane, 'The Destiny of Man. I: The Golden Age – and Then!'
57. Haldane, 'Auld Hornie, F.R.S.'; and Lewis, 'A Reply to Professor Haldane' in his *Of Other Worlds*, pp. 74–85. See Adams, 'Last Judgement', pp. 482–3.
58. Bernal, *The World, The Flesh and the Devil*, p. 20. On Bernal's work, see Goldsmith, *Sage*; and Swann and Aprahamian (eds), *J. D. Bernal*.
59. See Goldsmith, *Sage*, pp. 51–2.
60. Fournier d'Albe, *Quo Vadimus?* and *Hephaestos*.

61. Jennings, *Prometheus*.
62. Mitchell, *Hanno*, ch. 5. The book was somewhat tongue-in-cheek because it also included a chapter on exploring the centre of the Earth.
63. Hatfield, *Automaton*; see also his *The Inventor and His World*.
64. Low, *The Future* and *Our Wonderful World of Tomorrow*. The novel is *Mars Breaks Through*.
65. See Clark, *JBS*, p. 87.
66. Birkenhead, *The World in 2030 A.D.*, p. 152. *Daedalus*, like the rest of the 'Today and Tomorrow' series, was a small, pocket-sized book, while Birkenhead's was a more substantial affair with modernist illustrations.
67. Evans, *The World of To-morrow*. Evans had prepared the children's edition of Wells's *Outline of History*.
68. The 1938 series 'It Happened in 1963' is mentioned in Calder, *Hurtling Towards 2000 A.D.*, p. 4. On Calder's popular science writing, see Bowler, *Science for All*, pp. 207–8.
69. See, for instance, Kevles, *In the Name of Eugenics*, ch. 12.
70. See Krementsov, *Revolutionary Experiments: The Quest for Immortality in Bolshevik Science and Fiction*.
71. For a more general exploration of this theme, see: McLaren, *Reproduction by Design*; Armstrong, *Modernism, Technology and the Body*; and Chapter 11 below.
72. Chase, *Men and Machines*, p. 34. Millikan (like Rutherford in Britain) thought speculations about unlimited power from the atom were unfounded – see, for instance, his *Science and the New Civilization*, ch. 4.
73. Millikan, 'Science Lights the Torch'; and De Forest, 'Communication', chs 2 and 6 in Beard (ed.), *Toward Civilization*. Beard had edited an earlier volume, *Whither Mankind*, in which a number of intellectuals and moralists had expressed more negative opinions.
74. See, for instance, Franklin, 'America as Science Fiction: 1939', pp. 115–17.
75. Futuristic advertisements created by Arthur Radebaugh for the Bohn Aluminum and Brass Co. are reproduced in Hanks and Hoy, *American Streamlined Design* and in Sheller, *Aluminum Dreams*. For designs in plastic, see Meikle, *American Plastic*. More generally, see Bush, *The Streamlined Decade*; Meikle, *Twentieth-Century Limited*; and Smith, *Making the Modern*.
76. On the New York World's Fair, see: Gelenter, *1939: The Lost World of the Fair*; and Harrison, *Dawn of a New Day*. H. G. Wells visited the fair and wrote about it: see Kohn, 'Social Ideals in a World's Fair'. More generally on the impact of expositions, see: Rydell, *All the World's a Fair*; and the same author's *World of Fairs*; also Rydell and Schiavo, *Designing Tomorrow*.
77. See Ross, *Strange Weather*, esp. ch. 3, which notes how the enthusiasm of the individualists who wrote science fiction was overtaken by the corporate image of the future promoted at the World's Fair. The same theme emerges in Franklin, 'America as Science Fiction: 1939'.
78. Kaempffert published a second collection of his *Science Today and Tomorrow* in 1947. See also Brown, *The Challenge of Man's Future*; Calder (ed.), *The World in 1984* (collected from a *New Scientist* series of 1964); and Kahn and Wiener, *The Year 2000*. For further details, see the epilogue below.

79. Haldane, 'Biological Possibilities for the Human Species in the Next Ten Thousand Years'. This appears in a CIBA Foundation volume edited by Wolstenholme, *Man and His Future*.
80. See, for instance, Benford, 'Old Legends'.
81. Chapters 4 and 5 of Ross, *Strange Weather* outline the ideological implications of these events.

3 How We'll Live

1. The edition of London's *The Iron Heel* listed in the bibliography is the 12th of 1932. The book predicts socialist revolutions, initially repressed, but eventually triumphant.
2. Duhamel, *America the Menace*. On the complex reactions of French society to the approach of American-style culture, see: Panchasi, *Future Tense*; and Ross, *Fast Cars, Clean Bodies*.
3. See John Gloag's contribution to the 'Today and Tomorrow' series, *Artifex: Or the Future of Craftsmanship*.
4. The 'Today and Tomorrow' titles are Hatfield, *Automaton* and Garrett, *Ouroboros*. See also Birkenhead, *The World in 2030 A.D.*, p. 17; Langdon-Davies, *A Short History of the Future*, pp. 197–8; and Furnas, *The Next Hundred Years*, ch. 19. For a detailed analysis of the impact of mechanization, see Nye, *America's Assembly Line*. On Gernsback, see Cheng, *Astounding Wonder*, p. 195.
5. Low, *Our Wonderful World of Tomorrow*, ch. 19. See also *Practical Mechanics*, February 1934, front cover; 'Marvels of the Mechanical Man', pp. 208–9 and 218 and 'Marvels at the New York World's Fair', ibid., June 1939, p. 472.
6. Kahn and Wiener, *The Year 2000*, p. 94. Asimov's 'Satisfaction Guaranteed' is reprinted in his *Earth Is Room Enough*, pp. 104–16; on his robot stories, see Warrick, 'Ethical Evolving Artificial Intelligence'.
7. Chase, *Men and Machines*, ch. 8; see also Leonard, *Tools of Tomorrow*, ch. 5. Flanders, 'The New Age and the New Men' is in Beard (ed.), *Toward Civilization*, ch. 1, see pp. 20–30. See also Low, *The Future*, ch. 2, 'Sound and Silence'; Calder, *The Birth of the Future*, pp. 193–200; Teague, *Design This Day*, ch. 2; and Mumford, *Technics and Civilization*, pp. 5 and 423–8.
8. Birkenhead, *The World in 2030 A.D.*, p. 88.
9. De Forest, 'Communication', in Beard (ed.), *Toward Civilization*, ch. 6, see p. 135. See also Birkenhead, *The World in 2030 A.D.*, p. 91; Langdon-Davies, *A Short History of the Future*, p. 209; and Furnas, *The Next Hundred Years*, ch. 27. On the BBC's mission, see Reith, *Broadcast over Britain*. On the impact of radio and other new means of communication, see Chapter 5.
10. Figures given by Bryson, *One Summer*, p. 98.
11. *Conquest*, 1 (November 1919), p. 9.
12. Horrigan, 'The Home of Tomorrow'; also Nye, *Electrifying America*; and Gooday, *Domesticating Electricity*. On Tesla's emergence as a symbol of the new age, see Carlson, *Tesla: Inventor of the Electrical Age*.
13. Edison, 'The Woman of the Future', quoted in Gooday, *Domesticating Electricity*, p. 149.

14. E. Austin, 'Our Housemaid – Electricity: Recent Progress in Cooking and Heating', *Conquest*, 1 (January 1920): 112–18; Frank E. Perkins, 'Washing-Up "De-Lux"', ibid., p. 147. Christofleu, *Les dernières nouvautés de la science et de l'industrie*, pp. 100–8. Low, 'Women Must Invent', *Armchair Science*, 2 (March 1931): pp. 674–6; and on the vacuum cleaner *Our Wonderful World of Tomorrow*, p. 269. See also Michael Egan, 'Science in the Home: The All-Electric Home of the Future', *Armchair Science*, 1 (January 1930): 625–7.

15. Mee (ed.), *Harmsworth Popular Science* (1914 reprint), vol. 4, ch. 8; Low, *The Future*, ch. 3; and Gernsback, 'Fifty Years from Now'. See also Furnas, *The Next Hundred Years*, p. 21. For the *Popular Mechanics* reference, see Benford, *The Wonderful Future that Never Was*, p. 42. See also Gerald Carr, 'Fluorescence – the Lighting of the Future', *Armchair Science*, 9 (May 1937): 94; and Low, *Electronics Everywhere*, pp. 94–6.

16. Low, *Electronics Everywhere*, pp. 105–6. Thanks to Katy Price, 'Science is Golden' has been shown at several history of science meetings in Britain, but is not generally available. On the 'new kitchen', see Panchasi, *Future Tense*, ch. 1; and on the replicator Clarke, *Profiles of the Future*, ch. 13.

17. Benford, *The Wonderful Future that Never Was*, pp., 61, 67, 71–5. Low also called for mechanisms to sterilize clothes: see his *Our Wonderful World of Tomorrow*, ch. 15.

18. Low, *Our Wonderful World of Tomorrow*, ch. 21; and 'Artificial Life is Best', *Armchair Science*, 3 (April 1931): 6–8. On the early growth of the synthetics industry, see Bud, *The Uses of Life*, ch. 2.

19. Birnstingl, *Lares et Penates*, p. 33.

20. Benford (ed.), *The Wonderful Future that Never Was*, pp. 56–9; Kaempffert, *Science Today and Tomorrow*, ch. 9; and *Science Today and Tomorrow: Second Series*, ch. 3. For *The Graduate* quotation, see Meikle, *American Plastic*, p. 3. Meikle's *Twentieth-Century Limited* explores how the design potentials of the new materials were exploited, as does Hanks and Hoy's *American Streamlined Design*. The claim that many of these innovations were pioneered first by the aluminium industry is made by Sheller, *Aluminum Dreams*.

21. Frankl, *Form and Re-Form*. For historians' assessments of the movement in interior design, see: Bush, *The Streamlined Decade*, ch. 8; and Hanks and Hoy, *American Streamlined Design*, esp. the images on pp. 207–11.

22. Langdon-Davies, *A Short History of the Future*, p. 199.

23. Meikle, *American Plastic*, p. 154.

24. Birkenhead, *The World in 2030 A.D.*, p. 95; Low, 'Marvels of the Future', *Armchair Science*, 1 (August 1929): 214–18 and *The Future*, ch. 12; and Evans, *The World of To-morrow*, pp. 97–8.

25. Benford (ed.), *The Wonderful Future that Never Was*, pp. 50 and 54; and Alex Carlisle, 'Glass Clothes: Shall we all be wearing Glass?' *Armchair Science*, 5 (January 1934): 649–51.

26. On nylon, see Meikle, *American Plastic*, ch. 5.

27. On detergents, see Foertsch, *American Culture in the 1940s*, pp. 194–6; and on ultrasound, Benford (ed.), *The Wonderful Future that Never Was*, p. 60.

28. Lockhart-Mummery, *After Us*, ch. 6; Evans, *The World of To-Morrow*, ch. 7; and Langdon-Davies, *A Short History of the Future*, p. 221. Charles, *The Twilight of*

Parenthood, p. 9 quotes Sir John Russell on the productivity of scientific industry and agriculture and goes on to express concern about the declining population.

29. Calder, *Birth of the Future*, pp. 45 and 52; Benford (ed.), *The Wonderful Future that Never Was*, p. 67 and on the TV dinner (first actually produced in 1953), p. 65.

30. Haldane, *Daedalus*, pp. 38–9; and in Dronamraju (ed.), *Haldane's Daedalus Revisited*, p. 34; and Birkenhead, *The World in 2030 A.D.*, pp. 18–20. This was the sort of coincidence that led Haldane to accuse Birkenhead of plagiarism.

31. Lockhart-Mummery, *After Us*, pp. 78–81; Barklay, 'Sugar from Wood', *Armchair Science*, 6 (October 1934): 447–9; Forden, 'Sugar from Chalk?' ibid., 4 (December 1932): 568; and Low, 'What Hydrogenation Means to You', *Armchair Science*, 6 (May 1933): 74–7. See also 'Chemists Build a New World', ibid., 7 (March 1936): 694–7. On grass as potential food, see Benford (ed.), *The Wonderful Future that Never Was*, pp. 66–7.

32. Furnas, *The Next Hundred Years*, ch. 27; Lockhart-Mummery, *After Us*, pp. 86–7; and Evans, *The World of To-morrow*, p. 95. For the 'Today and Tomorrow' titles, see Hartley and Leyel, *Lucullus*, pp. 40–1; and Jones, *Hermes*, p. 56 and pp. 59–60.

33. Asimov, 'Visit to the World's Fair of 2014'. See also Brown, *The Challenge of Man's Future*, ch. 5; and Calder (ed.), *The World in 1984*, vol. I, pp. 16–17 and vol. II, pp. 181–4.

34. For the quotation, see Low, *Our Wonderful World of Tomorrow*, p. 269. See also his *The Future*, ch. 11; 'The Future of Women', *Armchair Science*, 1 (March 1930): 710–13; and 'Women Must Invent', ibid., 2 (March 1931): 674–6.

35. Evans, *The World of To-Morrow*, p. 142.

36. Birkenhead, *The World in 2030 A.D.*, ch. 8.

37. Langdon-Davies, *A Short History of the Future*, pp. 226–32 and for the quotation p. 254.

38. Anthony Ludovici's *Lysistrata* is an example of a conservative thinker denouncing the rise of women. We shall return to these wider issues in Chapter 11 below.

39. Birnstingl, *Lares et Penates*, ch. 3, see p. 46.

40. Cowan, *More Work for Mother*; see also Nye, *Electrifying America*, ch. 6.

4 Where We'll Live

1. Birnstingl, *Lares et Penates*, p. 56; for the quotation on substitutes, see Chapter 3 above, n. 19.

2. Low, *Our Wonderful World of Tomorrow*, pp. 263–5.

3. Birnstingl, *Lares et Penates*, pp. 58–9.

4. These predictions from *Popular Mechanics* are reproduced in Benford (ed.), *The Wonderful Future that Never Was*, pp. 39–47.

5. See Horrigan, 'The Home of Tomorrow'; and Corn and Horrigan, *Yesterday's Tomorrows*, pp. 72–7. On Fuller's designs, see Marks and Buckminster Fuller, *The Dymaxion World of Buckminster Fuller*, pp. 184–227. *Popular Mechanics'* account of Fuller's house is reproduced in Benford (ed.), *The Wonderful Future that Never Was*, p. 44.

6. Calder, *Birth of the Future*, ch. 10; see also Corn and Horrigan, *Yesterday's Tomorrows*, p. 84; and Asimov, 'Visit to the World's Fair of 2014'.

7. Edgell, *The American Architecture of To-Day*, pp. 73 and 84; for the Corbett report in *Popular Mechanics*, see Benford (ed.), *The Wonderful Future that Never Was*, p. 28 and on garbage disposal p. 32.
8. Ferris, *The Metropolis of Tomorrow*; see Currell, *American Culture in the 1920s*, ch. 4.
9. Fuller, *Atlantis*, pp. 29 and 92–3; and Lee, *Crowds*, chs 6 and 8. For a discussion of how new technologies created an interactive society, see Cassedy, *Connected*.
10. Le Corbusier, *The City of To-Morrow*. See Panchasi, *Future Tense*, pp. 62–74 and later sections of this chapter.
11. Examples from the 1920s are illustrated in Benford (ed.), *The Wonderful Future that Never Was*, pp. 22–7.
12. Reprinted in *The Short Stories of H. G. Wells*; see pp. 798–801 for a description of the cities.
13. *When the Sleeper Wakes*, later retitled *The Sleeper Awakes*; see ch. 5, 'The Moving Ways' for the best description of the city.
14. Lang's visit to New York was reported by *Popular Mechanics*; see Benford (ed.), *The Wonderful Future that Never Was*, p. 21.
15. For the *Commonweal* quotation, see Carter, *The Twenties in America*, p. 97; see also Ohana, *The Futurist Syndrome*.
16. Low, *Our Wonderful World of Tomorrow*, p. 266; and 'Where Shall We Live in 2031?' *Armchair Science*, 3 (June 1931): 134–6.
17. Mumford, *The Culture of Cities*, ch. 4.
18. Berkner, 'The Rise of Megalopolis' in Calder (ed.), *The World in 1984*, vol. 2, pp. 144–9; and Kahn and Wiener, *The Year 2000*, p. 61.
19. For the predictions in this paragraph, see Benford (ed.), *The Wonderful Future that Never Was*, pp. 30–2; and for a wider discussion of the expansion of the suburbs, Baxandall and Ewen, *Picture Windows*.
20. Teague, *Design This Day*, p. 212.
21. Originally published in French in 1924.
22. Barman, *Balbus*, pp. 56–8.
23. Le Corbusier, *The City of To-Morrow*, p. 6. The edition cited is a modern facsimile of the 1947 reprint, which was itself a facsimile of the 1929 original, translated from Le Corbusier's *Urbanisme* of 1924.
24. Lockhart-Mummery, *After Us*, ch. 11.
25. Calder, *Birth of the Future*, ch. 10. See also Evans, *The World of To-Morrow*, ch. 6, although Evans admitted that people didn't want to live in flats.
26. See Frank G. Novak, Jr., *Lewis Mumford and Patrick Geddes: The Correspondence*, 'Introduction'. On Geddes, see Kitchen, *A Most Unsettling Person*.
27. Howard, *Garden Cities of To-Morrow*; Mumford's essay in the new edition cited is pp. 29–40.
28. Lockhart-Mummery, *After Us*, frontispiece (see Fig. 4.2, p. 65).
29. On the early twentieth-century origins of modernism, see Pevsner, *Pioneers of Modern Design*; Banham, *Theory and Design in the First Machine Age*; and Ohama, *The Futurist Syndrome*. On the contrast between European and American design, see Greenhalgh, *Ephemeral Vistas*, ch. 6; Rydell, *World of Fairs*; and Rydell and Schiavo, *Designing Tomorrow*.
30. Frankl, *Form and Re-Form*, p. 47.

31. Lipmann is quoted in Gelenter, *The Lost World of the Fair*, p. 363. Gelenter also provides descriptions of the displays mentioned here, as do Köhlstedt, 'Utopia Realized' (and several other chapters in Joseph J. Corn's edited volume *Imagining Tomorrow*); see also Santomasso, 'The Design of Reason'. For Wells's visit, see Kohn, 'Social Ideals in a World's Fair'.
32. For Geddes's factory designs, see his *Horizons*, ch. 10. For Tottenville, see Benford, *The Wonderful Future that Never Was*, pp. 33–5.
33. See Berman, *All That Is Solid Melts into Air*, ch. 5. On the expressways, see Chapter 6 below.
34. On the role of scientific progress in the Festival, see Banham and Hillier (eds), *A Tonic to the Nation*; Forgan, 'Festivals of Science and the Two Cultures'; and Conekin, *'The Autobiography of a Nation'*, ch. 3. For the opposing views that the Festival also promoted nostalgia, see, for instance, Addison, *Now the War is Over*, pp. 208–9; and Hewison, *Culture and Consensus*, ch. 3.
35. See Morris, *The Best of Eagle*; and Morris and Hallwood, *Living with Eagles*.
36. Jeremiah, *Architecture and Design for the Family in Britain*, esp. pp. 124–42.

5 Communicating and Computing

1. Wells, *The Sleeper Awakes*, pp. 56–7; and for the quotation, his *The Work, Wealth and Happiness of Mankind*, pp. 152–3. On Gernsback, see Douglas, 'Amateur Operators and American Broadcasting', p. 36. For the *Popular Mechanics* items, see Benford (ed.), *The Wonderful Future that Never Was*, pp. 34–5 and 40.
2. Low, 'Telegraphing Photographs', *Armchair Science*, 1 (July 1929): 212–15. The other predictions are from *Popular Mechanics*, see Benford (ed.), *The Wonderful Future that Never Was*, pp. 89 and 95–6.
3. David Masters, 'Making Screen Artists Live: The Promise of Plastic Films', *Conquest*, 6 (October 1925): 479–80; and Gwendolyn Carlier, 'Cinema Pictures in Colour', *Conquest*, 5 (October 1924): 495–9.
4. Duhamel, *America the Menace*, p. 35.
5. Stapledon, *Star Maker*, ch. 3. The fact that the subjects were humanoid aliens hardly blurred the point Stapledon was trying to make.
6. Kaempffert, *Science Today and Tomorrow*, ch. 17 – the chapter title is 'Electric Immortality'.
7. Weightman, *Signor Marconi's Magic Box*, chs 34 and 39.
8. See, for instance, Douglas, 'Amateur Operators and American Broadcasting' and the same author's *Inventing American Broadcasting* and *Listening In*.
9. M. A. Laqui, 'Wireless Schemes of the Future', *Conquest*, 7 (March 1926): 133–4.
10. W. G. W. Mitchell, 'Wireless and the World's Work: Radio as an Aid to Navigation', *Conquest*, 5 (March 1924): 200–1.
11. Laqui, 'Distant Control by Wireless', *Conquest*, 6 (November 1925): 481–2; A. M. Low, 'Steering Aeroplanes by Wireless', *Armchair Science*, 2 (March 1930): 62–4; and the same author's *Wireless Possibilities*, ch. 5. Low's *The Future*, pp. 91–2 mentions the same possibility and predicts that power broadcasting will come eventually; on Tesla's work in this area, see Carlson, *Tesla: Inventor of the Electrical Age*, pp. 207–13 and ch. 15.

12. On the pre-war predictions, see Douglas, 'Amateur Operators and American Broadcasting', pp. 40–41.
13. Low, *Wireless Possibilities*, p. 35; 'On my Travels', *Armchair Science*, 1 (April 1929): 17–18; Evans, *The World of To-Morrow*, p. 33; and Furnas, *The Next Hundred Years*, p. 238.
14. Low, *Electronics Everywhere*, p. 165.
15. 'Speech and Music from the Ether', *Conquest*, 2 (June 1921): 309; and Benford (ed.), *The Wonderful Future that Never Was*, p. 87.
16. Birkenhead, *The World in 2030 A.D.*, pp. 114–15.
17. Gibbs, *The Day after Tomorrow*, pp. 31–2; Marconi's 1912 interview with *Technical World* is described by Weightman in his *Signor Marconi's Magic Box*, pp. 261–2.
18. Fournier d'Albe, 'The Future of Wireless', *Armchair Science*, 1 (July 1929): 225–6; E. W. 'Has Television a Future?' *Armchair Science*, 1 (January 1930): 625–6; and Evans, *The World of To-Morrow*, p. 33.
19. St Barbe Baker, 'What Wireless will Mean to Africa', *Conquest*, 5 (October 1924): 483–5; De Forest, 'Communication', p. 135; Compton, 'Physics and the Future', address delivered to the 'Science and Society' symposium in Ottawa 1938, quoted in Doug Russell, 'Popularization and the Challenge to Science-Centrism', pp. 42–3.
20. See Pegg, *Broadcasting and Society, 1918–1939*, p. 162.
21. Low, 'On my Travels', *Armchair Science*, 1 (March 1930): 727–9.
22. Leonard, *Tools of Tomorrow*, p. 268.
23. *Mars Breaks Through* was serialized as 'The Great Murchison Mystery' in *Armchair Science*, 1936–37. See also Low, *The Future*, ch. 21; and Evans, *The World of To-Morrow*, p. 34.
24. On the link between Lodge's interests in radio and spiritualism, see Douglas, *Listening In*, ch. 2. On Lodge's experiment with the BBC, see W. P. Jolly, *Sir Oliver Lodge*, p. 228.
25. Low, *Wireless Possibilities*, ch. 3; and De Forest in Benford (ed.), *The Wonderful Future that Never Was*, p. 91.
26. Bryson, *One Summer*, ch. 27, which also tells the story of how Filo T. Farnsworth developed the system subsequently exploited by RCA.
27. Quoted in Kamm and Baird, *John Logie Baird: A Life*, p. 184. See also the editorial 'Television: How it Works – Its Future Development', *Armchair Science*, 1 (April 1929): 40–2; H. Barton Chapple, 'Television – What It Offers Today', ibid., 3 (December 1931): 528–30; and for the quotation, Chapple, 'The Three Screens of Television', ibid., 3 (January 1932): 595–7, p. 597.
28. Calder, *The Birth of the Future*, ch. 9; and Low, *Electronics Everywhere*, p. 49; on the *Popular Mechanics* feature, see Benford (ed.), *The Wonderful Future that Never Was*, pp. 92–3.
29. Nicolas Marin, 'Où en est la télévision pratique?' *Science et Monde*, No. 182 (May 1935): 1064–7; and 'Anticipations', No. 186 (September 1935): 1346–7.
30. Furnas, *The Next Hundred Years*, ch. 23; Roy C. Norris, 'Look Out for Looking-In', *Armchair Science*, 8 (May 1936): 57 and 83; 'Practical Television', *Practical and Amateur Wireless*, 29 August 1937: 614 and photograph of German TV camera p. 602.
31. For the *Popular Mechanics* predictions, see Benford (ed.), *The Wonderful Future that Never Was*, pp. 94–5; Kaempffert, *Science Today and Tomorrow*, pp. 230–40;

Clarke, *The Exploration of Space*, pp. 156–7; and for his later reflections *Profiles of the Future*, ch. 16.

32. Barry, 'Mass Communication in 1984'.
33. See Paul Ceruzzi, 'An Unforseen Revolution: Computers and Expectations, 1935–1985'.
34. This point is made by Gary Westfahl in his introduction to Westfahl *et al.* (eds), *Science Fiction and the Prediction of the Future*, p. 3.
35. See Hampton and MacKay, 'The Internet and the Analogical Myths of Science Fiction'. A 'logic' was a home computer in terminology of the story.
36. Low, *Electronics Everywhere*, ch. 9; Berkeley, *Giant Brains*, ch. 12; and Benford (ed.), *The Wonderful Future that Never Was*, pp. 108 and 175.
37. 'Franchise', reprinted in Asimov, *Earth Is Room Enough*, pp. 53–66.
38. Wilkes, 'A World Dominated by Computers', which appeared in Calder's edited volume predicting the world of 1984; see also Calder, *Technopolis*, p. 19.
39. These reports are from *Popular Mechanics*; see Benford (ed.), *The Wonderful Future that Never Was*, pp. 72–5 and 100–1.
40. See Turner, *From Counterculture to Cyberculture*.

6 Getting Around

1. Fuller, *Pegasus: Problems of Transportation*.
2. See Carrington, *Rudyard Kipling*, p. 432.
3. For the American figure, see Bryson, *One Summer*, p. 98; for Britain, see Pugh, *We Danced All Night*, pp. 243–4; and more generally Thorold, *The Motoring Age*. On Huxley and motoring, see Murray, *Aldous Huxley*, p. 222. Le Corbusier's enthusiasm for the Paris traffic is in his *The City of To-Morrow*, p. 3.
4. W. Harold Johnson, 'Motor Fuels and the Future', *Conquest*, 1 (June 1920): 399–402 and on hydrogen p. 550; and Johnson, 'Electricity and the Motor Car', ibid. (September 1920): 515–20.
5. J. Harrison, 'The Possibility of a Diesel-Engined Car', *Armchair Science*, 2 (April 1930): 39–42; and 'Do we Build Cars the Wrong Way Round?', ibid., 2 (June 1930): 175–7. See also Norman Fuller, *Dymaxion Car: Buckminster Fuller*; and Marks and Buckminster Fuller, *The Dymaxion World of Buckmister Fuller*. In 2010, Norman Foster built a reconstruction of the Dymaxion car.
6. See Del Sesto, 'Wasn't the Future of Nuclear Energy Wonderful?', pp. 60–4. For more details on the emergence of the nuclear industry, see Chapter 10 below.
7. Benford, *The Wonderful Future that Never Was*, pp. 148–9.
8. See Banham, *Theory and Design in the First Machine Age*, p. 103.
9. See Meikle, *Twentieth-Century Limited*, pp. 140–1 and ch. 7; F. Rowlinson, 'A Real "Streamline" Car', *Conquest*, 3 (April 1922): 236–8; and Low, *The Future*, ch. 5, esp. the image facing p. 32.
10. For details and images of these futuristic designs, see Sheller, *Aluminum Dreams*; Hanks and Hoy, *American Streamlined Design*; and Bush, *The Streamlined Decade*. Geddes's reference to Blue Bird is in his *Horizons*, pp. 29 and 54–7. The *Popular Mechanics* designs (including the Vibarti bus) are reproduced in Benford, *The Wonderful Future that Never Was*, pp. 140–1 and 144–5. On Henry Ford's involvement, see, for instance, Meikle, 'Plastic, Material of a Thousand Uses', p. 90.

11. Eadhamite appears in Wells's 'A Story of the Days to Come': see his *Short Stories*, p. 818 and in *The Sleeper Awakes*, p. 126. See also Low, *Our Wonderful World of Tomorrow*, ch. 20; and 'Arterial Roads', *Armchair Science*, 2 (March 1931): 674–6; also 'Walking on Rubber Roads', *Tit-Bits*, 17 July 1920: 399; and R. E. Norman, 'And Next – Glass Roads?' *Armchair Science*, 5 (November 1933): 491–2.

12. Benford, *The Wonderful Future that Never Was*, pp. 80 and 143.

13. Le Corbusier, *The City of To-Morrow*, pp. 130–2.

14. Low, 'Motoring in 1983', *Armchair Science*, 5 (July 1933): 210–13; and Lockhart-Mummery, *After Us*, pp. 135–6.

15. Rollins, 'Whose Landscape? Technology, Fascism and Environmentalism on the National Socialist Autobahn'; Merriman, 'A Power for Good or Evil: Geographies of the M1 in Late Fifties Britain'; and Shand, 'The Reichsautobahn: Symbol for the Third Reich'.

16. Franklin, 'America as Science Fiction: 1939'; and Berman, *All That Is Solid Melts into Air*.

17. Joad and many others contributed to Williams-Ellis's edited volume *Britain and the Beast* in 1937.

18. Low, *The Future*, p. 79; and Lockhart-Mummery, *After Us*, p. 117.

19. 'The Roads Must Roll' is reprinted in Heinlein, *The Man Who Sold the Moon*, pp. 54–93.

20. David Masters, 'The Railway of Tomorrow', *Conquest*, 5 (March 1924): 338–40; the technology was already envisioned before the Great War. See Mee (ed.), *Harmsworth Popular Science*, vol. 3, pp. 1690–1.

21. Quoted in Horniman, *How to Make the Railways Pay for the War*, p. 114. The book includes a detailed discussion of the debates surrounding Gettie's proposal.

22. Martin, 'The Railways of Tomorrow'.

23. J. Clayton, 'The Locomotive – Its Development and Future', *Meccano Magazine*, 15 (September 1930): 688–91; E. S. P. Rawston, '100 m.p.h. Trains Are on the Way', *Armchair Science*, 6 (February 1935): 704–7; and Leonard, *Tools of Tomorrow*, p. 204.

24. Benford, *The Wonderful Future that Never Was*, p. 132; and Calder, *The Birth of the Future*, ch. 8.

25. Russell Thomas, 'All Electric Railway', *Armchair Science*, 3 (September 1931): 364–6; S. James, 'Locomotives of Tomorrow', ibid., 4 (September 1932): 381–2; Lt Col. Sir Gordon Hearn, 'Is Wholesale Electrification Possible?' ibid., 7 (October 1935): 428; and Sommerfeld, *Speed, Space and Time*, part 6, ch. 2.

26. Sommerfeld, *Speed, Space and Time*, part 6, ch. 3, see p. 262; F. J. Camm, 'Streamlined Trains', *Practical Mechanics*, July 1934: 446–8; and J. T. C. Moore-Brabazon, 'Shall We All Travel by Rail Car in the Future?', *Armchair Science*, 5 (March 1935): 762–3. On the American designs, see especially Bush, *The Streamlined Decade*, ch. 5.

27. Daniel Caire, 'Traction a vapeur et traction électrique', *L'Illustration*, No. 5128 (21 June 1941): 263–5; and Antoine Icart, 'Le chemin de fer n'a pas fini de nous étonner', *Jeunesse et Technique*, No. 2 (December 1961): 116–25.

28. Mee (ed.), *Harmsworth Popular Science*, vol. 3, pp. 168–89; and on *Popular Mechanics*, see Benford, *The Wonderful Future that Never Was*, pp. 134–6.

29. 'The Gyroscope and Mono-Rail Transport', *Meccano Magazine*, 18 (July 1933): 494–5.

30. Anon, 'Unusual Railways', *Practical Mechanics* (February 1935): 202–4; the system is also mentioned by Sommerfeld, *Speed, Space and Time*, pp. 264–5.

31. Benford, *The Wonderful Future that Never Was*, pp. 132–3, 136 and 137; and Kaempffert, *Science Today and Tomorrow*, p. 215.

32. *Meccano Magazine*, 15 (August 1930): front cover and 'High Speeds Transport by Overhead Railway', pp. 594–5 and 628; V. E. Johnson, 'Monorail Systems', *Practical Mechanics* (June 1935): 404–5 and 432; see also ibid. (February 1935): front cover and pp. 202–4. For the post-war plans, see Lawrence Dunn, 'Railway with the Speed of Aircraft', *Eagle Annual*, 1 (1951): 106–8. For the French expectations, see Serge Hyr, 'Destiné aux grandes villes: le "Cosmopolitan" fait ses débuts à la campagne', *Jeunesse et Technique*, No. 3 (January 1962): 220–30.

33. *Meccano Magazine*, 17 (December 1932): front cover and 'High Speed Trackless Railways', pp. 910–11 and 952.

34. See Benford, *The Wonderful Future that Never Was*, p. 138; and Antoine Icart, 'Une revolution en marche: le cousin d'air', *Jeunesse et Technique*, No. 5 (March 1962): 428–59.

35. e.g. in the 'Science Jottings' column in the *Illustrated London News*, 13 December 1913: 996.

36. Leonard Henslow, 'Tunnelling under the Channel', *Armchair Science*, 1 (February 1930): 662–4; 'The New Channel Tunnel', *Meccano Magazine*, 14 (February 1929): 130–2; and Frank Ellison, 'Bridging the English Channel: Famous Scientist's Amazing Scheme', ibid. (December 1929): 952–3; see also Evans, *The World of Tomorrow*, p. 45.

37. Pierre Devaux, 'A propos du projet de tunnel Europe–Afrique à Gibraltar', *Science et Monde*, No. 178 (January 1935): 814–16.

38. See Brosnan, *Future Tense*, p. 51.

39. Reported in *Popular Mechanics*; see Benford, *The Wonderful Future that Never Was*, pp. 136–8.

40. See Prof. M. J. Wise, 'How a Channel Tunnel (or Bridge) Would Change South-East England', *New Scientist*, 15 (16 August 1962): 392–4. Wise mentions the alternative of a bridge, but for more detail, see R. Reuter, 'Le Pont sur la Manche: realité de demain . . . ?' *Jeunesse et Technique*, No. 5 (March 1962): 460–8.

41. Leonard, *Tools of Tomorrow*, ch. 6.

42. Horniman, *How to Make the Railways Pay for the War*, pp. 240–2.

43. The 1909 proposals were reported in Mee (ed.), *Harmsworth Popular Science*, vol. 4, pp. 2320–1; see also A. B. Sawyer, 'A Great Water Transport Scheme', *Conquest*, 6 (May 1925): 262; and J. F. Pownall, 'Why Not an All-England Lockless Canal?' *Armchair Science*, 7 (June 1935): 178–81; also November 1935: 506.

44. Leonard, *Tools of Tomorrow*, ch. 6; and Sommerfeld, *Speed, Space and Time*, ch. 4.

45. See, for instance, anon., 'A Return to Wind Power: The Possibilities of the Rotor-Boat', *Conquest*, 6 (January 1925): 115–16. On diesel power, see Benford, *The Wonderful Future that Never Was*, pp. 128–31 and the section entitled 'An Engine of Revolution' in Mee (ed.), *Harmsworth Popular Science*, vol. 3, pp. 1915–31; also W. D. Horsnell, 'Ships of the Future', *Conquest*, 1 (March 1920):

46–66. On Turbo-electric power, see anon., 'Electric Propulsion for Ships', *Conquest*, 1 (March 1920): 240.

46. Acworth, *Back to the Coal Standard*; and on pulverized coal, George W. Greenland, 'The Ship of the Future', *Armchair Science*, 1 (December 1929): 525–6.

47. On nuclear ships, see, for instance, Calder, *Living with the Atom*, ch. 9.

48. Geddes, *Horizons*, pp. 36–41; see also Bush, *The Streamlined Decade*, pp. 46–7.

49. On the German plans, see Benford, *The Wonderful Future that Never Was*, pp. 130–1; and Low, 'Is the "Queen Mary" Out of Date?' *Armchair Science*, 8 (May 1936): 54–6.

50. Cockerell, 'The Prospects for Hover Transport'. On hydrofoils, see Benford, *The Wonderful Future that Never Was*, pp. 130–2.

7 Taking to the Air

1. See Edgerton, *England and the Aeroplane*, especially the preface. Like Edgerton's *Warfare State*, this book's insistence on the influence of the military on the development of aviation technology is linked to his claims about the extent to which the British Government was committed to spending on weapons technology in general. For an example of the view that the military applications perverted the natural development of aviation, see, for instance, Mackworth-Praed, *Aviation: The Pioneer Years*, e.g. p. 219.

2. 'The Argonauts of the Air' is reprinted in *The Short Stories of H. G. Wells*, pp. 391–405 – for the quotation, see p. 405. 'A Dream of Armageddon' is reprinted at pp. 1118–48 and 'A Story of the Days to Come' at pp. 796–897 – see p. 806 for the passenger flights. Chapter 6 of the 1910 version of *The Sleeper Awakes* describes the monoplanes and the flying stages from which they are launched, while ch. 25 is 'The Coming of the Aeroplanes'. Mr Butteridge's invention appears in the first chapter of *The War in the Air*.

3. 'With the Night Mail' is reprinted in Kipling's *Actions and Reactions*, pp. 111–67; and 'As Easy as A.B.C.' in *A Diversity of Creatures*, pp. 1–44.

4. For the Wells quotation and the *Observer* supplement, see: Gibbs-Smith, *Aviation*, p. 148; and Sampson, *Empires of the Sky*, p. 23. The images in Mee (ed.), *Harmsworth Popular Science* are on pp. 1325 and 3017. For the interwar predictions, see Chapter 9 below.

5. Grahame-White and Harper, *The Aeroplane*, ch. 19 and frontispiece. On Grahame-White's career as an aviation celebrity, see the biography by Graham Wallace.

6. *Daily Mirror*, 7 September 1929, front page and p. 9.

7. Low, *Our Wonderful World of Tomorrow*, ch. 4; Furnas, *The Next Hundred Years*, pp. 225–33; Leonard, *Tools of Tomorrow*, pp. 232–59; Birkenhead, *The World in 2030 A.D.*, ch. 6; Lockhart-Mummery, *After Us*, ch. 18, esp. pp. 204–6; and Sommerfeld, *Speed, Space and Time*, chs 6 and 7.

8. See, for instance, Swanwick, *Frankenstein and His Master: Aviation for World Service*.

9. Quoted in Wohl, *The Spectacle of Flight: Aviation and the Western Imagination, 1920–1950*, p. 49. This, and Wohl's companion book on the period 1908–1918, provide detail of the various countries' reaction to the prospects for aviation. See also Sampson, *Empires of the Sky*.

10. For the *New York Times* quote, see Bilstein, *Flight in America*, p. 39; and on the quasi-religious emotions aroused, Corn, *The Winged Gospel*. There are many books on Lindbergh and his influence, including the biographies by A. Scott Berg and Perry D. Luckett. On aviation in pulp magazines, see Cheng, *Astounding Wonder*, pp. 41–2. For the quotations, see: Guggenheim, *The Seven Skies*, p. 69; and Byrd, *Skyward*, pp. 311 and 325.

11. C. G. G., 'On the Great Flights', *The Aeroplane*, 32 (25 May 1927): 601–12, see p. 601; and Burney, *The World, the Air and the Future*.

12. Burney, *The World, The Air and the Future*, p. 343.

13. 'The Giant Aeroplane of the Future' (an interview with Handley Page), *Conquest*, 1 (November 1919): 26–30. On the limitations of early services, see: M. A. L., 'Commercial Aviation', ibid. (September 1920): 537; and R. A. de V. Robertson, 'Getting the World into the Air', ibid., 6 (June 1925): 303–7. Details of some early routes are given by M. B. Egan, 'Aeroplanes of Today', ibid., 2 (January 1921): 109–15. For a survey of early airline developments, see: Simpson, *Empires of the Sky*; and Mackworth-Praed, *Aviation: The Pioneer Years*.

14. On these developments, see, for instance, Gibbs-Smith, *Aviation*; and Sampson, *Empires of the Sky*. Shute's prediction is in a chapter included in Burney's *The World, the Air and the Future*, ch. 6, p. 279. See also Forbes-Smith (writing as the Master of Sempill), *The Air and the Plain Man*, p. 73 (note, however, that the fastest express trains could reach speeds equivalent to that quoted for the aircraft). For Geddes's designs, see his *Horizons*, ch. 5, which includes a prediction of speeds of up to 1,000 mph; see also Bush, *The Streamlined Decade*, pp. 25–40; and Hanks and Hoy, *American Streamlined Design*, pp. 223–9.

15. On searchlight beacons proposed for Europe, see Yves Tinin, 'Vols de nuit', *Science et Monde*, No. 184 (July 1935): 1191–4; and Christofleau, *Les Dernières nouveautés de la science et de l'industrie*, pp. 4–7. America developed a similar system, reported in 'The Lighting of Night Air Routes', *Meccano Magazine*, 15 (August 1930): 602–3. On radio beacons, see R. Quarendon, 'Safe Landing in Fog', *Discovery*, n.s. 1 (November 1938): 381–5. For Guggenheim's suggestions, see his *Seven Skies*, pp. 80–5 and 108–9; and on the German guidance system for bombers, see Johnson, *The Secret War*, ch. 1 (ch. 2 deals with the development of radar).

16. Proposals for overhead runways can be found in Barman, *Balbus*, pp. 56–8; Gernsback, 'Fifty Years from Now'; H. T. Winter, 'Airports of the Future', *Armchair Science*, 5 (February 1934): 730–72; 'Aviation of the Future: Aeroplane Landing Grounds in City Centres', *Meccano Magazine*, 17 (May 1932): 330–1 (also featured on the front cover); and from *Popular Mechanics* in Benford, *The Wonderful Future that Never Was*, pp. 120–1. The London scheme is described in 'King's Cross Aerodrome', *Flight*, 23 (12 June 1931): 523 and a later version in 'An Airport in the Heart of London', *Meccano Magazine*, 18 (May 1933): 340–1; see Barker and Hyde, *London as It Might Have Been*, p. 212. For the rotating airstrip, see 'City Airports of the Future', *Meccano Magazine*, 22 (February 1937): 65 and 70 (again featured on the front cover). Images of some of these schemes also appear in Corn, *The Winged Gospel*, pp. 70–7. On the 1946 plans for Paris and New York, see Charles Brachet, 'Aéroports Intercontinentaux', *Science et Vie*, No. 342 (March 1946): 99–106 and front cover.

17. Stewart, *Aeolus*, p. 35. For other calls to develop VTOL, see: F. J. Camm, 'Aircraft of the Future', *Practical Mechanics*, 1 (June 1934): 398–400; and Leonard, *Tools of Tomorrow*, pp. 245–7. The enthusiasts are Low, *Our Wonderful World of Tomorrow*, p. 54 and Lockhart-Mummery, *After Us*, p. 120. For concerns about viability, see, for instance: 'The Vertical Flight Problem', *Conquest*, 6 (November 1925): 505; and J. H. Crowe, 'What Hope has the Helicopter?' *Armchair Science*, 5 (July 1933): 231–3. For the *Popular Mechanics* prediction, see Benford, *The Wonderful Future that Never Was*, p. 154; and for a projected design for a twin-rotor helicopter airbus, see Hanks and Hoy, *American Streamlined Design*, p. 225.

18. Details of the German scheme can be found in Henri Le Masson, 'Europe-Amérique du sud: une realization Allemand', *Science et Monde*, No. 177 (December 1934): 761–3. On Armstrong's plan, see 'Wonderful Seadrome Scheme: Aeroplane Landing Grounds in the Atlantic', *Meccano Magazine*, 14 (January 1929): 2–3 and image on front cover; also Norman J. Hulbert, 'Transatlantic Air Routes of the Future', *Armchair Science*, 1 (February 1930): 655–7; see also S. H. Jones, 'First Atlantic Seadrome for Seaplanes', ibid., 5 (December 1933): 581–2. On the movies, see: Brosnan, *Future Tense*, pp. 45–6, which includes a still image; and Hunter, *British Science Fiction Cinema*, pp. 27–8.

19. On the Clippers, see: Bilstein, *Flight in America*, ch. 3; and Bush, *The Streamlined Decade*, pp. 25–40. On the British flying boats, see Sims, *Adventurous Empires*. On the *Mercury/Maia* combination, see T. R. Robinson, 'The Short-Mayo Composite Aircraft', *Meccano Magazine*, 23 (June 1938): 320–1; and for the French reaction, C. Rougeron, 'L'Avenir de l'Avion Porte-Avions', *La Science et la Vie*, 53 (June 1938): 455–61. For futuristic designs, see, for instance, Hanks and Hoy, *American Streamlined Design*, pp. 223–9; Geddes, *Horizons*, pp. 108–21; and Benford, *The Wonderful Future that Never Was*, p. 157. For the freight-carrying flying boat, see John Towers, 'Air Freighters of the Future', *Practical Mechanics*, 9 (September 1942): 340–2 and front cover image.

20. Sinclair, *Airships in Peace and War*, p. 286. In addition to Spanner's *Gentlemen Prefer Aeroplanes!*, see his *This Airship Business* and *The Tragedy of the 'R 101'*. There are several general books on the airship programmes, including: Jackson, *Airships in Peace and War*; and Payne, *Lighter than Air*.

21. Low's comment (written when Britain had already abandoned its airship programme) is in his 'Germany's Monster Airships', *Armchair Science*, 7 (June 1935): 150–2. For an adventure story involving an imaginary new gas, see Westerman, *The Airship 'Golden Hind'*. The 1929 image is reproduced in Benford, *The Wonderful Future that Never Was*, p. 191; see also W. D. Verschoyle, 'Electro-Gravitic Lift', *Practical Mechanics*, 9 (February 1942): 138; and Campbell, *The Atomic Story*, p. 277.

22. 'The Royal Mail Dirigible of the Future', *Illustrated London News*, 134 (16 January 1909): 97. For popular anticipations, see, for instance, George Whale, 'The Future of the Commercial Airship', *Conquest*, 1 (November 1919): 353–6 and his article on the same theme in *Discovery*, 2 (January 1921): 18–19; also T. W. Blake, 'A Transatlantic Airship Service', *Discovery*, 4 (September 1923): 227–30. Burney's idea is in his *The World, the Air and the Future*, pp. 237–46.

23. *Daily Express*, 6 October 1930: 10; Shute's autobiography is his *Slide Rule*, see pp. 54–106 on airships.

24. There are numerous coffee-table books about the Zeppelins, of which the most useful is Archbold, *Hindenburg*, actually a wider survey of the airship industry; for a more academic account of the programme and its wider influence, see Syon, *Zeppelin!* For the proposal of an *entente* between America and Europe, see Henri Bouchet's report, 'La Fin du "Hindenburg"', *L'Illustration*, No. 4915 (15 May 1937): 61–4.

25. Some of these fears are explored by M. J. Bernard Davy in his *Air Power and Civilization*. Orwell's critique is in his 'As I Please' article of 12 May 1944, reprinted in his *Collected Essays*, vol. 3, pp. 173–6. Major L. H. Bauer's work on the problem of centrifugal forces at high speeds was reported in 'Flight – 1,000 Miles an Hour: Planes Could Do It, Men Could Not', *Popular Science Siftings*, 65 (4 March 1924): 421; Harley's talk was reproduced in *The Listener*, 18 (3 November 1937): 949–51.

26. There has been some debate about the identity of 'Neon', but I have been informed by Bernard Ackworth's son, the Revd Richard Ackworth, that Marion Ackworth was his father's sister-in-law. Sueter's response is his *Airmen or Noahs* and Pollen's article is his 'Is Air Power a Delusion?' *Discovery*, 8 (April 1927): 123–7.

27. Ackworth, *The Navies of Today and Tomorrow*, p. 120. His opposition to technical experts is noted by Edgerton, *Warfare State*, pp. 117–18. I have described his religious views in my *Reconciling Science and Religion*, pp. 293–5.

28. For details of these programmes, see Corn, *The Winged Gospel*, ch. 5. See also Huxley, *Brave New World*, e.g. p. 45.

29. The prejudice against autogyros is noted by Frances Jones, 'Let's Fly Home To-morrow', *Armchair Science*, 4 (December 1932): 596–9; other articles in the magazine on the topic include: Low, 'Roof to Roof Flying', ibid., 7 (October 1936): 402 and the Earl of Cardigan, 'Mr. Everyman's Aeroplane', ibid., 8 (February 1937): 492–3. For the *Popular Mechanics* features, see Benford, *The Wonderful Future that Never Was*, pp. 124–5 and 148–54. See also Teague, *Design this Day*, p. 241; and Asimov, 'Personal Flight 2000 A.D.' reprinted in his *Today and Tomorrow and . . .*, pp. 211–16.

30. See Wilson, *Where's My Jetpack?*

31. The fallacies are listed in Westfahl *et al.* (eds), *Science Fiction and the Prediction of the Future*, pp. 9–22.

32. Space forbids detailed listing of the vast literature on modern aviation, but on the hopes and disappointments of the British aircraft industry, see Hamilton-Patterson, *Empire of the Clouds*. Nevil Shute's novel *No Highway* imagined events paralleling the Comet disasters in 1948, just as the design of the plane was taking shape.

33. On the German work, see Myhra, *Secret Aircraft Designs of the Third Reich*. Lean's film and the events it was based on are described by Hamilton-Patterson; see also Kaempffert, *Explorations in Science*, ch. 5.

34. Sartre, 'Travelling by Air in 1984'; see Kaempffert, *Science Today and Tomorrow*, p. 217; and Ian Bruce, 'Aircraft that Changes Shape in Flight', *Boy's Own Paper*, 83 (January 1961): 24–5.

35. Anon., 'Cool towards the Supersonic Airliner', *New Scientist*, 15 (27 September 1962): 661.

36. There are many books on Concorde. One of the most recent, Jonathan Glancey's *Concorde: The Rise and Fall of the Supersonic Airliner*, has some information about earlier and rival projects; see chs 1–4.
37. For details, see Del Sesto, 'Wasn't the Future of Nuclear Energy Wonderful?' pp. 64–8. The *Popular Mechanics* prediction is in Benford, *The Wonderful Future that Never Was*, pp. 160–2; the *Eagle* image is reproduced in Tatarsky (ed.), *Eagle Annual: The 1950s*, pp. 120–1. See also Campbell, *The Atomic Story*, p. 163.
38. See Evans, *Greenglow and the Search for Gravity Control* (Project Greenglow was BAE Systems' attempt to develop anti-gravity). For Clarke's predictions, see his *Profiles of the Future*, chs 6 and 7.

8 Journey into Space

1. Mitchell, *Hanno*, p. 84; and editorial in *Daily Mirror*, 9 September 1929: 9.
2. Westfahl *et al.* (eds), *Science Fiction and the Prediction of the Future*, pp. 9–22.
3. See Winter, *Prelude to the Space Age*. An earlier but still useful account is Williams and Epstein, *The Rocket Pioneers*. On the pioneers' self-creation of their identity, see Geppert, 'Space *Personae*'.
4. The 1919 paper is reprinted in Goddard and Pendray (eds), *The Papers of Robert H. Goddard*, vol. 1, pp. 337–406. See also Goddard, 'That Moon Rocket Proposition', *Scientific American*, 124 (26 February 1921): 166. For a popular biography, see Dewey, *Robert Goddard: Space Pioneer*. On the 1924 proposal, see: Hugh Pollard, 'Professor Goddard's Rocket to the Moon', *Discovery*, 5 (June 1924): 90–2; and Frank T. Addyman, 'To the Moon: Modern Projects for Reaching our Satellite', *Conquest*, 6 (December 1924): 76. On the Russian reactions, see Siddiqi, *The Red Rockets' Glare*, pp. 64–5 and 87.
5. Lasser, *The Conquest of Space*, ch. 14 on exotic power sources and pp. 238–9 on Mars. Information on Lasser and the American Interplanetary Society can be found in Cheng, *Astounding Wonder*, ch. 5; Kilgore, *Astrofuturism*, ch. 1; and McCurdy, *Space and the American Imagination*, ch. 1.
6. On the Rocketport, see Gelenter, *1939: The Lost World of the Fair*, p. 155; Bush, *The Streamlined Decade*, pp. 166–7; and Benford, *The Wonderful Future that Never Was*, p. 170.
7. Cleator, *Rockets through Space*, on anti-gravity see pp. 133–4 and on Mars pp. 204–6. For his later comment on the reaction, see his *Into Space*, p. 7. Reactions to Wooley's later comment were based on a misunderstanding (noted below). In addition to those cited in n. 4, popular science articles on the topic include C. S., 'Will Man Migrate to the Moon?' *Armchair Science*, 1 (November 1929): 550; Frank Bardon, 'Our Son's Sons May "Rocket" to the Moon', ibid., 2 (August 1930): 300–1; Low, 'The Higher the Fewer', ibid., 3 (July 1931): 198–200; Cleator, 'World Rocketry Today', ibid., 4 (August 1934): 286–7; and Ley and Cleator, 'The Rocket Controversy', ibid., 7 (April 1935): 17–19. See also Max Vallier, 'Can We Fly to the Stars?' *Discovery*, 9 (July 1928): 210–12; Anon., 'American Views on Rocket Flying', ibid., 9 (September 1928): 291–3; J. G. Strong, 'Are Rockets to Mars a Possibility?' ibid., 16 (February 1935): 42–4; and Charles E. Philp, 'Is Rocketry Progressing?' ibid., 18 (September 1937): 269–71 and 277. Evans's book is his *The World of To-Morrow*, see ch. 5.

8. On the Russian situation, see Siddiqi, *The Red Rockets' Glare*; Smith, *Rockets and Revolution*; and Maurer *et al.* (eds), *Soviet Space Culture*.
9. Ley's *Rockets and Space Travel* is a useful English-language account of the work in Germany. On the early origins of interest in the topic, see Brandau, 'Cultivating the Cosmos: Spaceflight Thought in Imperial Germany'.
10. See Neufeld, *Von Braun: Dreamer of Space, Engineer of War*.
11. Gary Westfahl stresses that much of the pulp science fiction was not very 'hard', see his *Cosmic Engineers*, pp. 62–3. On the ideology of the privately financed but heroic engineer, see Ross, *Strange Weather*, ch. 3. On the different styles of Clarke's writing, see, for instance, Olander and Greenberg (eds), *Arthur C. Clarke*.
12. Bernal, *The World, the Flesh and the Devil*, pp. 20–34. Clarke's hope of a technology 'Beyond Gravity' is expressed in his *Profiles of the Future*, ch. 5. In 1956, Lloyd Malan reported that the Glen L. Martin company was investigating anti-gravity, see his *Men, Rockets and Space*, p. 267. Juvenile literature includes Appleton II, *Tom Swift in the Race to the Moon* (there are many other titles in the series, not all of which deal with space) and Johns, *Kings of Space* and *Return to Mars*.
13. Space forbids proper coverage of this issue, but for a detailed analysis, see Stephen J. Dick, *The Biological Universe*. For details of the panic created by Welles's broadcast, see Cantril, *The Invasion from Mars*, which also includes the programme's script.
14. For Harper's article, see Carter, *The Creation of Tomorrow*, p. 29. For Haldane's 'The Last Judgment', see his *Possible Worlds*, pp. 287–312; and for Stapledon, *Last and First Men*, pp. 223–33.
15. Stills from Lang's movie appeared, for instance, in the *Illustrated London News* (24 August 1929): 346–7; see also the article (2 November 1929): 701. On the response to the Wells movie, see Stover, *The Prophetic Soul*.
16. See Orwell's 'Boys' Weeklies' in his *Collected Essays, Journalism and Letters*, vol. 1, pp. 460–85; and Richards's response at pp. 486–93. 'The War in Space' was launched in the *Boy's Magazine*, 8 (24 July 1926). On Flash Gordon, see Kohl, 'Flash Gordon Conquers the Great Depression'.
17. The radio series was turned into a book, Chilton, *Journey into Space*, which was widely translated (the French version is *Mission dans l'éspace*). There were several follow-up series.
18. *Destination Moon* is covered in Warren, *Keep Watching the Skies!*, 1, pp. 2–6 and Quatermass by Hutchings, 'We're All Martians Now'. On Dan Dare, see Watkins, 'Piloting the Nation'; and on his creator, Frank Hampton, see Crompton, *The Man who Drew Tomorrow*. On *The Eagle* more generally, see Morris (ed.), *The Best of Eagle*; and Tatarsky, *Eagle Annual: The 1950s*; also Chapman, *British Comics: A Cultural History*, ch. 2. On Morris's moral crusade, see Morris and Hallwood, *Living with Eagles*. Johns's science fiction was rather far-fetched, but some of Moore's books offered a more down-to-earth vision of exploring Mars; see Johns, *Kings of Space* and Moore, *Mission to Mars*.
19. Lewis, *The Cosmic Trilogy*, p. 20. For his views on science fiction, see the letters quoted in Green and Hooper, *C. S. Lewis: A Biography*, pp. 163 and 166–7. On his interaction with Clarke, see Poole, 'The Challenge of the Spaceship'.

20. Clarke, 'The Problem of Dr. Campbell' and 'When Will the Real Space Race Begin?' reprinted in his *Greetings, Carbon-Based Bipeds!*, pp. 48–53 and 492–5. On Woolley's comments, see also John Rudge's letter, *New Scientist*, 147 (16 September 1995): 52.

21. Details of the Soviet programme are derived from Siddiqi, *The Red Rockets' Glare* and Maurer *et al.* (eds), *Soviet Space Culture*. For Calder's comment, see his 'The Red Moon', *New Statesman*, 54 (12 October 1957): 452.

22. In addition to the books by Ley and von Braun, see Malan, *Men, Rockets and Space* and on the space station Kaempffert, *Explorations in Science*, ch. 1. Further details may be found in McCurdy, *Space and the American Imagination* and Kilgore, *Astrofuturism*. Many images from this period are reproduced in David Meerman Scott and Richard Jurek, *Marketing the Moon*. On Bonestell's imagery as reflecting America's manifest destiny, see Sage, 'Framing Space'.

23. John Wyndham's *The Outward Urge* was written with technical advice from Lucas Parkes. *Practical Mechanics* featured space vehicles in November 1951 and March 1954. See also E. Colston Shepherd, 'A Space Patroller for the R.A.F.', *New Scientist*, 15 (13 September 1962): 556–8. For the 'Ladybird' books, see: Carey, *How It Works: The Rocket*; and Worvill, *Exploring Space*. See also: William R. Macaulay, 'Crafting the Future'; and Farry and Kirby, 'The Universe Will be Televised'. More generally on European reactions, see Geppert (ed.), *Imagining Outer Space*.

24. Clarke's essays on geosynchronous satellites and on the challenge of the spaceship are reprinted in modified form in his *Greetings, Carbon-Based Bipeds!*, pp. 19–25 and 33–47. See also his *The Exploration of Space*, pp. 156–7 on the satellites. For studies of his influence, see Poole, 'The Challenge of the Spaceship'; Neil McAleer, *Odyssey*; and Olander and Greensberg (eds), *Arthur C. Clarke*.

25. The figures for the polls are derived from McCurdy, *Space and the American Imagination*, pp. 29 and 47. Heinlein's future history chart is reproduced in his *The Man who Sold the Moon*, front- and rear-end papers; see also James Gifford, *Robert A. Heinlein*, pp. 14–16. Clarke's predictions were in his novel *Prelude to Space*; for his later comments, see his *Greetings, Carbon-Based Bipeds!*, pp. 492–5. See also: Wyndham and Parkes, *The Outward Urge*; and Asimov, 'Visit to the World's Fair of 2014'.

26. Clarke's comments are in the reprinted version of his 'The Challenge of the Spaceship', *Greetings, Carbon-Based Bipeds!*, pp. 19–25. Expectations of atomic rockets can be found throughout Heinlein's early stories, also in Ley, *Rockets and Space Travel*, p. 283; Moore, *The Boys' Book of Space*, p. 117; Harper, *Dawn of the Space Age*, pp. x–xi; Malan, *Men, Rockets and Space*, p. 266; and (with warnings about the difficulties) in Cleator, *Into Space*, pp. 110–13. *Popular Mechanics* predicted atomic rockets in 1951 and 1958 – see Benford, *The Wonderful Future that Never Was*, pp. 162–5. See also Gerald W. Englert, 'Towards Thermonuclear Spacecraft Propulsion', *New Scientist*, 16 (4 October 1962): 16–18.

27. On the later development of the space programme, see Kilgore, *Astrofuturism*, ch. 5; McCurdy, *Space and the American Imagination*, chs 4–8; Scott and Jurek, *Marketing the Moon*; Klerkx, *Lost in Space*; and W. Patrick McCray, *The Visioneers*.

9 War

1. Nichols, *Cry Havoc!*, ch. 1; and Lefebure, *Scientific Disarmament*, p. 243. On American attitudes, including Edison's contributions, see Franklin, *War Stars*; and on Britain, Edgerton, *Warfare State*, ch. 3.
2. Low, 'Frightfulness and Humbug', *Armchair Science*, 4 (January 1933): 618–22; see also his *The Future*, ch. 15 and *Our Wonderful World of Tomorrow*, ch. 10. See Nichols, 'If They Liked, Scientists Could Save the World from War', *Armchair Science*, 4 (June 1932): 155–6. Haldane's contribution and the future war novels are discussed in more detail later in this chapter.
3. 'The Land Ironclads' is reprinted in *The Short Stories of H. G. Wells*, pp. 131–57. On the development of the tank and the debates on how it should be employed, see J. P. Harris, *Men, Ideas and Tanks*.
4. Lefebure's prediction of the sub-machine gun is in his *Scientific Disarmament*, p. 73. The quotation from Fuller is taken from his *The Reformation of War*, p. 168. For Basil Liddell Hart, see his *Paris: Or the Future of War* and *When Britain Goes to War*. On his mistaken belief in his influence on the Germans, see Alex Danchev, *Alchemist of War*, pp. 224–38. De Gaulle's *The Army of the Future* is a translation of his *Vers l'armée de métier* of 1934; on French military thinking, see Panchasi, *Future Tense*, ch. 3.
5. For the *Popular Mechanics* items, see Benford, *The Amazing Weapons that Never Were*, pp. 112–13. Novels entitled *The Flying Submarine* were published by Percy F. Westerman in 1912 and E. van Pedroe Savidge in 1922. In addition to Norton's *The Vanishing Fleets*, see Moffett, *The Conquest of America*.
6. Kenworthy's *New Wars: New Weapons* was published in 1930 and Charlton's *War from the Air* in 1935. Bywater's *Sea-Power in the Pacific* was revised for the new edition in 1934. On the future war novels mentioned here and in the rest of this chapter, see: Clarke, *Voices Prophesying War*; and Ceadel, 'Popular Fiction and the Next War, 1918–39'.
7. Lanchester, *Aircraft in Warfare*, p. 165; Grahame-White and Harper, *Air Power*, pp. 43–6; and Douhet, *The Command of the Air*, p. 22. The claim that flying is evil is from Charlton, *War from the Air*, p. 5. Recent studies include: Haapamaki, *The Coming of the Aerial War*; Holman, *The Next War in the Air*; and Peden, *Arms, Economics and British Strategy*. An older study is Bialer, *The Shadow of the Bomber*. On the emergence of the logic of deterrence, see Quester, *Deterrence before Hiroshima*.
8. The assumption that civilian planes could be used as bombers is explored in Holman, 'The Shadow of the Airliner'; see also Shute, *Slide Rule*, pp. 206 and 212. An early prediction of attack by radio-controlled planes is in Christofleau, *Les dernières nouveautés de la science et de l'industrie*, pp. 82–3; see also Charlton, *War from the Air*, pp. 165–73; Kenworthy, *New Wars: New Weapons*, pp. 115–16; and on American ideas, Franklin, *War Stars*, ch. 5. Low's aerial forts are described in his *Our Wonderful World of Tomorrow*, p. 126. For Shute's idea, see his *What Happened to the Corbetts*, pp. 113–18.
9. Macmillan, *Winds of Change*, p. 575.
10. Turner, *Britain's Air Peril*, p. 104. See also: Maxwell Deane, 'Horror in the Air', *Armchair Science*, 5 (December 1933): 567–9; and Low, 'The Devil in a Gas Mask', ibid., 7 (July 1935): 216–19; for *Popular Mechanics* articles on gas, see

Benford, *The Amazing Weapons That Never Were*, pp. 32 and 146–8. Gas and germs are described in Churchill's article, 'Shall We All Commit Suicide?' reprinted in his *Thoughts and Adventures*, pp. 174–81.

11. Hart, *Paris*, p. 51. Freeth's opinion is cited in S. H. Jones, 'Laying the Gas War Bogey', *Armchair Science*, 5 (March 1934): 774–5; see also Low, 'Frightfulness and Humbug', ibid., 4 (January 1933): 618–22 – this is two years earlier than the more pessimistic article cited in n. 10.

12. Details of the novels warning of poison gas are given in the bibliography; in addition to those named in the text, see also Neil Bell, Leslie Beresford, Ladbroke Black, Sarah Campion, Moray Dalton, Shaw Desmond, Frank Fawcett, George Godwin, Martin Hussingtree, Frank McIlraith and Roy Connolly, Joseph O'Neill, Simpson Stokes and S. Andrew Wood.

13. Marvell, *Three Men Make a World*. See also H. G. P. Castellain, 'Plagues by 'Plane', *Armchair Science*, 7 (March 1935): 758–62. J. T. Connington was the pen name of Alfred Walter Stewart, one-time Professor of Chemistry as Queen's University, Belfast.

14. Anon., 'The Light Ray that May Stop Wars: Actual Experiments', *Illustrated London News* (26 April 1924): 733–5. M. A. Laqui, 'Death-Rays and Moonshine', *Conquest*, 5 (July 1923): 382–3; also '"Death Rays" and "Heat Rays"', *Discovery*, 5 (June 1924): 83–4; and for *Popular Mechanics* coverage, see Benford, *The Amazing Weapons that Never Were*, pp. 36–7. The *Punch* cartoon from 4 June 1924 is reproduced in Campbell, *Rutherford*, p. 395, where it is noted that a New Zealand scientist also claimed to have developed a death ray. On Matthews, see: Barwell, *Death Ray Man*; and Foster, *The Death Ray*.

15. Low, 'The Truth about Death-Rays', *Armchair Science*, 2 (April 1930): 6–8; see also his 'Frightfulness and Humbug', ibid., 4 (January 1933): 618–22; and 'Horrors of Science', ibid., 7 (March 1935): 750–62. Churchill mentioned death rays in his 'Shall We All Commit Suicide?' reprinted in his *Thoughts and Adventures*, pp. 174–81, p. 178.

16. Carlson, *Tesla: Inventor of the Electrical Age*, pp. 381–9. For Stapledon's prediction, see his *Last and First Men*, pp. 26–31 – the ray then becomes part of the mythology of the next civilization as the 'Gordelupus mystery' – see pp. 64–5.

17. On the early speculations, see Weart, *Nuclear Fear*, part 1 and *The Rise of Nuclear Fear*, ch. 1 and 2; on Soddy, see Merricks, *The World Made New*, pp. 36 and 56–7. For the quotation, see Soddy, *The Interpretation of Radium*, p. 244; Wells's *The World Set Free* begins with a dedication acknowledging Soddy's influence.

18. For examples, see the Weart citations in n. 17 and for Churchill's speculations, see his 'Shall We All Commit Suicide?' and 'Fifty Years Hence', reprinted in *Thoughts and Adventures*, pp. 174–81 and 193–204, discussed in Farmello, *Churchill's Bomb*, pp. 30–1, 42–3 and 87. Rutherford's suggestion that splitting the atom might blow up the whole planet was reported by W. C. D. Whetham, 'Matter and Electricity', *Quarterly Review*, 199 (January 1904): 100–26, see p. 126. The idea that novae result from such explosions was mentioned in Kramers and Holst, *The Atom and the Bohr Theory of Its Structure*, p. 103 and in Leonard's *Tools of Tomorrow*, p. 54; for the late example, see Maurice-E. Nahemias, 'Vers l'utilisation de l'energie atomique: la rupture explosive de l'uranium', *Science et la Vie*, 62

(March 1940): 238–45. The *Punch* cartoon from 15 October 1924 is reproduced in Campbell, *Rutherford*, p. 407.

19. The text of the play is Nichols and Browne, *Wings over Europe*. See Carpenter, *Dramatists and the Bomb*, pp. 19–26; the New York production was publicized even in Britain – see the photograph reproduced in the *Illustrated London News* (5 January 1929): facing p. 1. On the science fiction coverage of the bomb, see Franklin, *War Stars*, ch. 8.

20. C. P. Snow, 'A New Means of Destruction?' *Discovery*, n.s. 1 (September 1939): 443–4; and, for the American reports, see La Follette, *Science on the Air*, pp. 186–7.

21. Laurence, *Dawn over Zero*, pp. 240–1; Arnold's views are reported in Dietz, *Atomic Energy Now and Tomorrow*, pp. 156–62; and for Campbell's echoing of these warnings, see his *The Atomic Story*, ch. 16. Asimov's admission about the ignorance of the dangers of radioactivity came in the 'Afterword' added to later printings of *Pebble in the Sky*, dated November 1982. For further examples, see the later chapters of Weart's surveys cited above.

22. Kahn, *On Thermonuclear War*, pp. 40–95 and 512–22. On fallout, see Devine, *Blowing in the Wind*; and on calls for disarmament, Wittner, *Confronting the Bomb*. British responses are discussed in a special issue of the *British Journal for the History of Science*, 45 (December 2012).

10 Energy and Environment

1. Ross, *Standing Room Only*, p. 114. On Crookes and Soddy, see Thaddeus J. Trenn, 'The Central Role of Energy in Soddy's Holistic and Critical Approach to Nuclear Science, Economics, and Social Responsibility'. See also Carr-Saunders, *Population*, pp. 52–3; East, *Mankind at the Crossroads*, pp. 161–2; and Knibbs, *The Shadow of the World's Future*, pp. 38–44. For further discussion of the population problem, see Chapter 12 below.

2. Haldane, *Daedalus*, pp. 24–5; Dronamraju (ed.), *Haldane's Daedalus Revisited*, pp. 29–30; Gibbs, *The Day after Tomorrow*, p. 23; Russell, *The Scientific Outlook*, pp. 153–4; Low, *Our Wonderful World of Tomorrow*, p. 42; Calder, *Birth of the Future*, p. 104; Furnas, *The Next Hundred Years*, pp. 185–6; and Lockhart-Mummery, *After Us*, pp. 196–7. Manning's *The Man Who Awoke* depicts the collapse of civilization 5,000 years in the future. On Hubbert and the debate over peak oil, see Inman, *The Oracle of Oil*; and for the 1964 prediction, H. W. Slotboom, 'Plentitude for Petroleum' in Calder (ed.), *The World in 1984*.

3. Anon., 'The Great Power Question: Harnessing Natural Energy', *Conquest*, 5 (October 1924): 500; and Kaempffert, *Science Today and Tomorrow*, ch. 8. On geothermal power, see Mee (ed.), *Harmsworth Popular Science*, vol. 6, pp. 4334–5; and Furnas, *The Next Hundred Years*, pp. 194–7.

4. See Rollo Appleyard, *Charles Parsons: His Life and Work*, pp. 234–8; Mee (ed.), *Harmsworth Popular Science*, vol. 6, pp. 4327–8; Leonard, *Tools of Tomorrow*, pp. 39–43; and Lockhart-Mummery, *After Us*, pp. 197–8.

5. On the Egyptian scheme, see Calder, *Birth of the Future*, p. 103; and I. O. Evans, 'Inventions of the Future', *Meccano Magazine*, 22 (August 1937): 452–3. See also Low, *Our Wonderful World of Tomorrow*, pp. 39–40; Leonard, *Tools of Tomorrow*,

pp. 46–8; and Furnas, *The Next Hundred Years*, pp. 191–5. On the Cabot-funded research at MIT, see Kaempffert, *Explorations in Science*, ch. 8.

6. On the French proposals, see Christofleau, *Les dernières nouveautés de la science et de l'industrie*, pp. 205–6. In 'A Story of the Days to Come' the Wind-Vane and Waterfall Trust generates all the world's electricity, see *The Short Stories of H. G. Wells*, pp. 800 and 820; also *The Sleeper Awakes*, p. 72. For Haldane's prediction, see his *Daedalus*, pp. 24–5; and Dronamraju (ed.), *Haldane's Daedalus Revisited*, pp. 30–1. For negative opinions, see: Low, *Our Wonderful World of Tomorrow*, p. 38; and Furnas, *The Next Hundred Years*, p. 190.

7. Arthur R. Burrows, 'Is Niagara Doomed?' *Conquest*, 2 (January 1921): 103–8; Anon., 'The Wonders of White Coal', *Tit-Bits* (17 July 1920): 400; Gordon Stokes, 'Harnessing the Clouds', *Conquest*, 1 (June 1920): 363–6; Edward C. Rashleigh, 'The Greatest Water Power in the World', *Discovery*, 12 (September 1931): 295–7; and Denys Parsons, 'Harnessing Victoria Falls', ibid., n.s. 1 (July 1938): 189–95.

8. Evans, *The World of To-Morrow*, facing p. 26. Haldane, 'The Last Judgement', in his *Possible Worlds*: 287–312, see pp. 294–5. For details of proposed tidal schemes, see Calder, *Birth of the Future*, p. 103; Furnas, *The Next Hundred Years*, p. 189 (on the Maine scheme); and Christofleau, *Les dernières nouveautés de la science et de l'industrie*, pp. 212–13 (on Mont Saint Michel). On the Straits of Gibraltar scheme, see Lockhart-Mummery, *After Us*, pp. 196–7. Magazine coverage included the *Harmsworth Popular Science*, vol. 6, pp. 4339–40; P. J. Risdon, 'The Severn Scheme: Pros and Cons of a Gigantic Project', *Conquest*, 2 (March 1921): 215–18; and Anon., 'The River Severn Barrage Scheme', *Discovery*, 14 (April 1933): 29–30.

9. Soddy, *The Interpretation of Radium*, pp. 164–73 and 229–50. On Soddy's other efforts to promote the promise of atomic energy, see: Merricks, *The World Made New*, pp. 52–66; and Trenn, 'The Central Role of Energy in Soddy's Holistic and Critical Approach to Nuclear Science, Economics, and Social Responsibility'.

10. On Le Bon, see Mary Jo Nye, 'Gustav Le Bon's Black Light: A Study in Physics and Philosophy in France at the Turn of the Century', *Historical Studies in the Physical Sciences*, 4 (1974): 163–95. On Rutherford's position, see David Wilson, *Rutherford; Simple Genius*, pp. 387–8, 467–9, 493 and 572–82; and on reports in America, Weart, *The Rise of Nuclear Fear*, p. 20. For Millikan's skepticism, see his *Science and the New Civilization*, pp. 111–12 and on his earlier views Kargon, *The Rise of Robert Millikan*. The 'Today and Tomorrow' item is Whyte, *Archimedes: Or the Future of Physics*.

11. Gibbs, *The Day After Tomorrow*, p. 29. For negative views, see Calder, *The Birth of the Future*, pp. 104–5; Lockhart-Mummery, *After Us*, p. 199; Kaempffert, *Science Today and Tomorrow*, pp. 120–22; also Evans, *The World of To-Morrow*, p. 24. See also Low, *Our Wonderful World of Tomorrow*, p. 45; Furnas, *The Next Hundred Years*, ch. 15; John Thomson, 'Using the Energy of the Atom in Industry', *Discovery*, 15 (May 1934): 142; and Leonard, *Tools of Tomorrow*, pp. 50–7.

12. Mee (ed.), *Harmsworth Popular Science*, vol. 1, pp. 129–37, see pp. 135–6; Anon., 'World's Most Terrible Secret', *Tit-Bits*, (26 June 1920): 362; Anon., 'Seeking to Disrupt the Atom: Immeasurable Energy', *Illustrated London News* (11 October 1924): 67–8; and Anon., 'The Electron's Hidden Magic: Will Science Harness its Mighty Power?' *Tit-Bits* (7 March 1925): 32. On the publicity surrounding the

splitting of the atom in 1932, see, for instance, Cathcart, *The Fly in the Cathedral*, ch. 14; also Crowther, 'Breaking Up the Atom', see p. 81; and R. J. Regnants, 'Splitting the Atom', *Armchair Science*, 4 (June 1932): 175–6. American comments on nuclear energy in this period are covered in, for instance, Weart, *Nuclear Fear*, part 1; and Nelson, *The Age of Radiance*.

13. J. J. Connington was the *nom de plume* of A. W. Stewart, one-time professor of chemistry at The Queen's University of Belfast. See *News Chronicle* (7 August 1945): 1 and 4. For details of the American responses mentioned here and in the following paragraph, see Weart, *Nuclear Fear*, parts 2 and 3; Boyer, *By the Bomb's Early Light*, part 4; Nelson, *The Age of Radiance*; and Del Sesto, 'Wasn't the Future of Nuclear Energy Wonderful?' On Britain, see Forgan, 'Atoms in Wonderland' and the collection of articles in the special issue of the *British Journal for the History of Science*, 45 (December 2012): 4. For Calder, see his *Technopolis*, p. 17.

14. Campbell, *The Atomic Story*, p. 261; Gamow, *Atomic Energy in Cosmic and Human Life*, pp. 151–9; and Blackett, *Military and Political Consequences of Atomic Energy*, pp. 93–8.

15. Titterton, *Facing the Atomic Future*, ch. 15; Jay, *Calder Hall*, ch. 6; and Brown, *The Challenge of Man's Future*, pp. 172–8. On the Calder Hall incident and the resulting enquiry (which didn't happen until the 1970s), see: Wynne, *Rationality and Ritual*; and more generally Mahaffey, *Atomic Accidents*.

16. Calder, *Living with the Atom*, pp. 208 and 228–30. On the link between the nuclear disarmament movement and environmentalism in Britain, see Burkett, 'The Campaign for Nuclear Disarmament'.

17. Huxley, *Memories*, p. 289; and Clark, *The Huxleys*, p. 285. See also Milo D. Nordyke, 'Excavation by Nuclear Explosions', *New Scientist*, 15 (5 July 1962): 36–40. On Teller, see Fleming, *Fixing the Sky*, ch. 7.

18. For details of these schemes, see Fleming, *Fixing the Sky*; and for the *Collier's* image, p. 175. See also Evans, *The World of To-Morrow*, pp. 91–3.

19. De Vries, *Species and Varieties*, p. 688; and for details of his reception, Endersby, 'Mutant Utopias'.

20. On MacDougal, see Kingsland, 'The Battling Botanist'; and more generally Campos, *Radium and the Secret of Life*. The later developments are covered in detail by Curry's *Evolution Made to Order.*

21. Russell, 'That the Earth Shall Produce More Food', *The Listener*, 18 (10 November 1937): 999–1001. Harold J. Shepstone, 'Beef of the Future', *Armchair Science*, 1 (November 1929): 468–9; and Furnas, *The Next Hundred Years*, chs 8 and 9. For hydroponics, see, e.g., Calder, *The Birth of the Future*, ch. 12.

22. The classic account is Joravsky, *The Lysenko Affair.* For the suggestion that Lysenko's work may have had some real value, see Graham, *Lysenko's Ghost.* There is an enormous literature on the rise of genetics; for my own survey, see Bowler, *The Mendelian Revolution.*

23. On Carson and the impact of *Silent Spring*, see Davis, *Banned.*

24. The quotation from Sauer's article in the *Journal of Farm Economics* (1938) is from Worster, *Dust Bowl*, p. 207; see also Sears, *Deserts on the March*, p. 157.

25. Anker, *Imperial Ecology*, pp. 110–17; and Wells *et al., The Science of Life*, Book 6, ch. 6. For Sear's reference to Wells, see *Deserts on the March*, p. 177. There is

a substantial secondary literature on the origins of ecology; for my own survey, see Bowler, *The Fontana/Norton History of the Environmental Sciences*.
26. Brown, *The Challenge of Man's Future*, pp. 142–3; and Revelle, 'A Long View from the Beach'. See Weart, *The Discovery of Global Warming* for other details.

11 Human Nature

1. A good example of a novel featuring germ warfare is Thurlow Craig's *Plague over London*, based on fears circulated in 1936 (see the footnote on p. 13). See also Chadwick, *The Death Guard*; and on the legacy of Shelley's imagery, John Turney, *Frankenstein's Footsteps*. On the influence of Loeb, see Pauly, *Controlling Life*; and for my own account of the resistance to materialism in the early twentieth century, see Bowler, *Reconciling Science and Religion*.
2. Ross, *Standing Room Only*, ch. 4. For more on the population issue, see below.
3. Calder, *The Conquest of Suffering*, p. xiii; on Mellanby, see pp. 15–16. See also Mee (ed.), *Harmsworth Popular Science*, group 12: 'Eugenics: the Ennobling of the Life of the Future'; and Wells *et al.*, *The Science of Life*, Book 7, ch. 4 (pp. 653–60 in the edition cited). On Saleeby, see Rodwell, 'Dr. Caleb Williams Saleeby: The Complete Eugenist'. On the publicity for medical science, see Hansen, *Picturing Medical Progress from Pasteur to Polio*. Note, however, that popular science magazines, at least in Britain, were only slowly beginning to treat this as a major topic of interest: see Bowler, 'Discovering Science from an Armchair'.
4. For the photograph, see *Conquest*, 2 (August 1921): 436; and for more details, see Nancy Knight, '"The New Light": X Rays and Medical Futurism'; on radium, see Louis A. Campos, *Radium and the Secret of Life*. For surveys, see Low, *The Future*, ch. 13 and *Our Wonderful World of Tomorrow*, ch. 11; Furnas, *The Next Hundred Years*, chs 3–6; Gibbs, *The Day after Tomorrow*, ch. 2; and Evans, *The World of To-Morrow*, ch. 10. See also: R. Cecil Owen, 'Vitamines [sic]: the Mighty Atoms of the Diet', *Conquest*, 2 (May 1921): 315–19; and Low, 'The Health Guard Will Advance', *Armchair Science*, 7 (August 1935): 274–7.
5. On Huxley, see Witkowski, 'Julian Huxley in the Laboratory'. See also Thomas Stephenson, 'The Conquests of Medical Science, II: The Hormones or Chemical Messengers of the Blood', *Conquest*, 2 (December 1920): 59–61; Jones, *Hermes*, p. 80; and Furnas, *The Next Hundred Years*, p. 30.
6. On Voronoff, see Hamilton, *The Monkey Gland Affair*; and more generally on these techniques, Armstrong, *Modernism, Technology and the Body*, ch. 5; McLaren, *Reproduction by Design*, esp. ch. 1; and Sengoopta, *The Most Secret Quintessence of Life*, ch. 3. These works give details of the reports in the American press; on Germany, see Paul Weindling, *Health, Race and German Politics*, esp. ch. 5; and on Russia, Kremenstov, *Revolutionary Experiments*. British reports include Theodor Robin, 'The Real Doctor Voronoff', *Armchair Science*, 1 (July 1929): 202–3; and 'Dr. Voronoff's Super-Sheep', ibid. (September 1929): 333–4; Low, 'Some Events in Our Children's Lives', ibid., 328–30; Voronoff, 'Longer Life from Gland-Grafting', ibid., 3 (October 1931): 407–9; and 'With Monkeys I Hope to Conquer Cancer', ibid., 5 (January 1934): 638–40; also E. E. Fournier d'Albe, 'Can Life Be Prolonged Indefinitely?' ibid., 3 (November 1931): 457–8.

7. Voronoff, *The Conquest of Life*, pp. 178–9; Kammerer, *Rejuvenation*, p. vii and chs 29–35; and Berman, *The Glands Regulating Personality*, ch. 13. The classic, but biased, account of Kammerer's Lamarckian experiments is Koestler, *The Case of the Midwife Toad*; but see also Bowler, *The Eclipse of Darwinism*, pp. 92–102.

8. Schmidt, *The Conquest of Old Age*, pp. 297–8. Huxley's 'Tissue-Culture King' was published in the *Cornhill Magazine*, 60 (1926): 422–57.

9. On Vincent and his debate with enthusiast Ernest H. Starling, see Long Hall, 'The Critic and the Advocate'. For the text of the BBC talk, see Hamilton Hartridge, 'Do You Want to Live Longer?' *The Listener*, 18 (17 November 1937): 1074–6; and for Kaempffert's chapter, see his *Science Today and Tomorrow, 2nd Series*, ch. 16. On Niehans, see McGrady, *The Youth Doctor*.

10. See Russell, *Hypatia*, pp. 68 and 78–9; Brittain, *Halcyon*, quotation from p. 91; Ludovici, *Lysistrata*, pp. 95–6; Schiller, *Tantalus*, pp. 57 and 67; and Jennings, *Prometheus*, pp. 90–1. For a survey of these debates, see, for instance, McLaren, *Reproduction by Design*, ch. 1.

11. Low, 'The Future of Women', *Armchair Science*, 1 (March 1930): 710–13; and 'Women Must Invent', ibid., 2 (March 1931): 674–6; see also his *The Future*, ch. 14; and *Our Wonderful World of Tomorrow*, ch. 9. For Watson's views, see his 'After the Family – What?'

12. The interactions between Haldane and the Huxleys are noted in Clark, *JBS*, pp. 56–7; more generally, see Adams, 'Last Judgement: The Visionary Biology of J. B. S. Haldane'; and Bud, *The Uses of Life*, ch. 3. See Huxley, *Chrome Yellow*, pp. 28 and 129–31; and Haldane, *Daedalus*, pp. 63–4 (in Dronamraju (ed.), *Haldane's Daedalus Revisited*, pp. 42–3). See also Haldane, 'The Destiny of Man', *Evening Standard* (10 October 1927): 7 and 9.

13. Russell, *The Scientific Outlook*, ch. 16; and Birkenhead, *The World in 2030 A.D.*, p. 16. On the reception of *Brave New World*, see, for instance, Watt (ed.), *Aldous Huxley: The Critical Heritage*, pp. 197–221.

14. On streamlining, see Cogdell, *Eugenic Design*, esp. ch. 1. See MacFie, *Metanthropos*, pp. 89–91; and Langdon-Davies, *A Short History of the Future*, pp. 254–6. Haldane's 'The Last Judgement' is in his *Possible Worlds*, pp. 287–312; Bernal's claim that ectogenesis offers more scope than eugenics is in *The World, the Flesh and the Devil*, pp. 38–9; for the press response to the book, see Goldsmith, *Sage*, pp. 51–2. See also Stapledon, *Last and First Men*, ch. 11.

15. Rostand, *Can Man Be Modified?*, pp. 80–5; and Haldane, 'Biological Possibilities for the Human Species in the Next Ten Thousand Years'. On Pincus and contemporary developments, see Pauly, *Controlling Life*, esp. pp. 177–83 and 191–4. Taylor's succinct list of predictions is in his *The Biological Time Bomb*, p. 218; for a wider discussion of this period, see Turney, *Frankenstein's Footsteps*, ch. 7.

16. Carrell, *Man the Unknown*, pp. 124–5. On Lodge and spiritualism, see Janet Oppenheim, *The Other World*, ch. 8; and more generally on the opposition to materialism, Bowler, *Reconciling Science and Religion*. See Bennett, *Apollonius*; and Gibbs, *The Day after Tomorrow*, ch. 4; Low's support for telepathy can be seen in his *The Future*, ch. 21 and *Our Wonderful World of Tomorrow*, ch. 16. His *Mars Breaks Through* was originally serialized in

Armchair Science under the title 'The Great Murchison Mystery'. On parapsychology and academic psychology, see Roger Smith, *The Fontana/ Norton History of the Human Sciences*, pp. 673–80.

17. Wells *et al., The Science of Life*, Book 8, ch. 4; Low, *Our Wonderful World of Tomorrow*, p. 150; and 'The Health Guard Will Advance', *Armchair Science*, 7 (August 1935): 274–7. Wright's story 'Brain' is in his *The New Gods Lead*, pp. 59–104; see also Kaempffert, *Science Today and Tomorrow, Second Series*, ch. 18; and Rostand, *Can Man Be Modified?*, pp. 57–60.

18. For Huxley's contacts with the eugenics movement, see, for instance, Murray, *Aldous Huxley*, pp. 274–5 (his brother Julian was also a convert). For a broad account of the movement, see Kevles, *In the Name of Eugenics*; and for a survey of recent literature, Alison Bashford and Philippa Levine (eds), *The Oxford Handbook to the History of Eugenics*.

19. Birkenhead, *The World in 2030 A.D.*, p. 14; and Carrell, *Man the Unknown*, pp. 300–1 and 318–19.

20. Macaulay, *What Not*, see pp. 10–13 and 144–5; Wright's story 'P.N. 40' is in his *The New Gods Lead*, pp. 149–93. For the Haldane quotation, see: *Daedalus*, pp. 40–1; and Dronamraju (ed.), *Haldane's Daedalus Revisited*, p. 35. See also Wells *et al., The Science of Life*, Book 9, ch. 2, esp. p. 879; Lockhart-Mummery, *After Us*, ch. 4; Muller, *Out of the Night*, pp. 137–41; and Russell, *The Scientific Outlook*, ch. 16 and pp. 175–6.

21. Armstrong, *The Survival of the Unfittest*, ch. 6. For the *New York Times* feature and other examples in American culture, see Cogdell, *Eugenic Design*, ch. 2; also Currell and Cogdell (eds), *Popular Eugenics*. On the hopes of some environmentalists, see McLaren, *Reproduction by Design*, ch. 6. See also: Schiller, *Tantalus*, pp. 61–3; and Blacker, *Birth Control and the State*, pp. 87–8.

22. Langdon-Davies imagined Britain's population being reduced to a tenth of its current level and whole areas of America returned to wilderness: see *A Short History of the Future*, pp. 170–1.

23. Vogt, *Road to Survival*, p. 211.

24. Darwin followed the new consensus that there were no significant genetic differences between the races – see *The Next Million Years*, pp. 79–80; but for the comment about the imaginary African historian, see *The Problems of World Population*, pp. 40–1. On these developments generally, see Bashford, *Global Population*, ch. 10.

25. Rostand's *Can Man Be Modified?* and Taylor's *Biological Time Bomb* are discussed above; Lessing's *DNA: At the Core of Life Itself* appeared in 1966. For details, see Turney, *Frankenstein's Footsteps*, ch. 7, which is entitled 'Priming the Biological Time Bomb'.

12 Epilogue

1. On the rise of futurology in the Cold War era, see, for instance, Ross, *Strange Weather*, ch. 5. For suggestions as to why prognostication often goes wrong, see Gary Westfall's 'Pitfalls of Prophecy' in Westfall *et al.* (eds), *Science Fiction and the Prediction of the Future*, pp. 9–22.

2. See Kahn and Wiener, *The Year 2000*, esp. ch. 11, and their 'The Next Thirty-Three Years', pp. 710–16. The latter is a contribution to the special issue 'Toward the Year 2000: Work in Progress', *Daedalus*, 96, no. 3 (1967). See also Calder (ed.), *The World in 1984*; and Asimov, 'Visit to the World's Fair of 2014'. On the recognition that computers could be liberating, see Fred Turner, *From Counterculture to Cyberculture*. For the Clarke predictions mentioned, see his *Profiles of the Future*, chs 5 and 6; and on O'Neill's space colonies, De Witt Kilgrove, *Astrofuturism*, ch. 5; and W. Patrick McCray, *The Visioneers*.

3. Anon., 'Science in Disrepute', *New Scientist*, 15 (6 September 1962): 489. 'The Tyranny of Technology' is the last chapter in Lapp's *The New Priesthood*. The suggestion that the invisible nature of the new threats was significant is in Commoner, *Science and Survival*, pp. 114–15.

4. Gabor, *Inventing the Future*, ch. 6, esp. pp. 75–6.

5. Calder, *Technopolis*, pp. 348–53. 'Mugwump' has a different and more specific meaning in the United States.

Bibliography

This listing contains all books and substantial articles consulted, both primary and secondary. It does not include newspaper articles or short pieces in magazines, for which sources are indicated in the footnotes.
Where reprinted editions are cited, the original date of publication (if known) is given in square brackets.

Ackworth, Bernard. *This Bondage: A Study of the 'Migration' of Birds, Insects and Aircraft, with Some Reflections on 'Evolution' and Relativity*. London: John Murray, 1929.
   The Navies of Today and Tomorrow: A Study of the Naval Crisis from Within. London: Eyre and Spottiswood, 1930.
   Back to the Coal Standard: The Future of Transport and Power. London: Eyre and Spottiswood, 1932.
Ackworth, Marion W. See 'Neon'.
Adams, Mark B. 'Last Judgement: The Visionary Biology of J. B. S. Haldane.' *Journal of the History of Biology*, 33 (2000): 457–91.
Addison, Hugh. *The Battle of London*. London: Herbert Jenkins, 1924.
Addison, Paul. *Now the War is Over: A Social History of Britain, 1945–51*. London: BBC/Jonathan Cape, 1985.
Akin, William E. *Technocracy and the American Dream: The Technocratic Movement, 1900–1941*. Berkeley, CA: University of California Press, 1997.
Aldiss, Brian and David Wingrove. *Billion Year Spree: The History of Science Fiction*. London: Paladin, 1986.
Aldridge, Alexandra. *The Scientific World View in Dystopia*. Ann Arbor, MI: UMI Research Press, 1984.
Anker, Peder. *Imperial Ecology: Environmental Order in the British Empire, 1895–1945*. Cambridge, MA: Harvard University Press, 2001.
Appleton, Victor II. *Tom Swift in the Race to the Moon*. London: Collins, 1969.
Appleyard, Rollo. *Charles Parsons: His Life and Work*. London: Constable, 1933.
Arapostathis, Stathis and Graeme Gooday. *Patently Contestable: Electric Technologies and Inventor Identities on Trial in Britain*. Cambridge, MA: MIT Press, 2013.
Archbold, Rick. *Hindenburg: An Illustrated History*. London: Weidenfeld & Nicolson, 1994.
Arlen, Michael. *Man's Mortality*. London: Heinemann, 1933.
Armstrong, Charles Wicksteed. *The Survival of the Unfittest*. London: C. W. Daniel, 1927.

Armstrong, Tim. *Modernism, Technology and the Body: A Cultural Study*. Cambridge University Press, 1998.

Ashley, Michael (ed.). *The History of the Science Fiction Magazine: Part 1*. London: New English Library, 1975.

(ed.). *The History of the Science Fiction Magazine: Part 2*. London: New English Library, 1975.

Asimov, Isaac. 'Visit to the World's Fair of 2014', *New York Times*, 16 August 1964, available at http://www.nytimes.com/books/97/03/23/lifetimes/asi-v-fair.html (accessed 22 February 2016).

Foundation. London: Panther, 1960 [1951].

Earth Is Room Enough. London: Panther, 1962 [1957].

The Martian Way and Other Stories. London: Panther, 1965.

Is Anyone There? London: Rapp and Whiting, 1968.

Today and Tomorrow and . . . London: Abelard-Schuman, 1974.

Towards Tomorrow. London: Coronet, 1974.

Asimov on Science Fiction. London: Granada, 1983.

Pebble in the Sky. New York: Ballantine Books, 1983. [1950]

Living in the Future. London: New English Library, 1985.

I, Asimov: A Memoir. New York: Doubleday, 1994.

The Caves of Steel. Reprinted London: Harper Collins, 1997 [1954].

I, Robot. Reprinted London: Harper Voyager, 2013 [1950].

Astor, John Jacob. *A Journey in Other Worlds: A Romance of the Future*. London: Longmans, Green & Co., 1894.

Atherton, Gertrude. *Black Oxen*. London: John Murray, 1923.

Austin, F. Britten. *The War-God Walks Again*. London: Williams & Norgate, 1926.

Banham, Mary and Bevis Hillier (eds). *A Tonic to the Nation: The Festival of Britain, 1951*. London: Thames and Hudson, 1976.

Banham, Rayner. *Theory and Design in the First Machine Age*. London: Architectural Press, 1960.

Bankoff, George. *The Conquest of Cancer*. London: Macdonald, 1947.

Barker, Felix and Ralph Hyde. *London as It Might Have Been*. London: John Murray, 1995.

Barman, Christian. *Balbus: Or the Future of Architecture*. London: Kegan Paul, 1927.

Barry, Sir Gerald. 'Mass Communication in 1984' in Nigel Calder (ed.), *The World in 1984*. London: Penguin, 1964, vol. 1, pp. 157–60.

Barwell, Ernest. *Death Ray Man: The Biography of Grindell Matthews, Inventor and Pioneer*. London: Hutchinson, n.d.

Bashford, Alison. *Global Population: History, Geopolitics, and Life on Earth*. New York: Columbia University Press, 2014.

and Philippa Levine (eds). *The Oxford Handbook to the History of Eugenics*. Oxford University Press, 2010.

Baxandall, Rosalyn and Elizabeth Ewen. *Picture Windows: How the Suburbs Happened*. New York: Basic Books, 2000.

Baxter, John. *Science Fiction in the Cinema*. New York: A. S. Barnes/London: A. Zwimmer, 1970.

Beard, Charles A. *Whither Mankind? A Panorama of Modern Civilization*. London: Longman Green, 1928.

(ed.). *Towards Civilization*. London: Longman Green, 1930.

Bell, Neil [Stephen Southwold]. *The Gas War of 1940*. London: Collins, 1941. [Originally published as *Valiant Clay*, 1931.]

Benford, Gregory. 'Old Legends' in Greg Bear (ed.), *New Legends*. London: Legend Books, 1996, pp. 292–306.

and the editors of *Popular Mectanics* (eds). *The Wonderful Future that Never Was*. New York: Hearst Books, 2010.

and the editors of *Popular Mechanics* (eds). *The Amazing Weapons that Never Were*. New York: Hearst Books, 2010.

Bennett, E. N. *Apollonius: Or the Present and Future of Psychical Research*. London: Kegan Paul, 1927

Berg, A. Scott. *Lindbergh*. London: Macmillan, 1998.

Bergonzi, Bernard. *The Early H. G. Wells*. Manchester University Press, 1961.

Berkeley, Edmund C. *Giant Brains or Machines That Think*. New York: Wiley/London: Chapman & Hall, 1949.

Berman, Louis. *The Glands Regulating Personality: A Study of the Glands of Internal Secretion in Relation to the Types of Human Nature*. 2nd edn, revised. New York: Macmillan, 1930.

Berman, Marshall. *All That Is Solid Melts into Air: The Experience of Modernity*. New edn. London: Verso, 2010 [1982].

Bernal, J. D. *The World, the Flesh and the Devil: An Enquiry into the Future of the Three Enemies of the Rational Soul*. London: Kegan Paul, 1929.

Bialer, Uri. *The Shadow of the Bomber: The Fear of Air Attack and British Politics, 1932–1939*. London: Royal Historical Society, 1980.

Bilstein, Roger E. *Flight in America, 1900–1983: From the Wrights to the Astronauts*. Baltimore, MD: Johns Hopkins University Press, 1984.

Birkenhead, the Earl of. *The World in 2030 A.D.* London: Hodder & Stoughton, 1930.

Birnstingl, H. J. *Lares et Penates: Or the Home of the Future*. London: Kegan Paul, 1928.

Black, Ladbroke. *The Poison War*. London: Stanley Paul, 1933.

Blacker, C. P. *Birth Control and the State: A Plea and a Forecast*. London: Kegan Paul, 1926.

Blackett, P. M. S. *Military and Political Consequences of Atomic Energy*. London: Turnstile Press, 1948.

Blom, Philip. *The Vertigo Years: Change and Culture in the West, 1900–1914*. London: Weidenfeld & Nicolson, 2008.

Bloom, Ursula. *He Lit the Lamp: A Biography of Professor A. M. Low*. Introd. Lord Brabazon of Tara. London: Burke, 1959.

Boon, Timothy. *Films of Fact: A History of Science in Documentary Films and Television*. London: Science Museum, 2007.

Borgstrom, Georg. *The Hungry Planet: The Modern World at the Edge of Famine*. New York: Collier Macmillan, 1965

Bowler, Peter J. *The Eclipse of Darwinism: Anti-Darwinian Evolution Theories in the Decades around 1900*. Baltimore, MD: Johns Hopkins University Press, 1983.

The Mendelian Revolution: The Emergence of Hereditarian Concepts in Modern Science and Society. London: Athlone, 1989.

The Fontana/Norton History of the Environmental Sciences. London: Fontana/ New York: Norton, 1992.

Reconciling Science and Religion: The Debate in Early Twentieth-Century Britain. University of Chicago Press, 2001.

Monkey Trials and Gorilla Sermons: Evolution and Christianity from Darwin to Intelligent Design. Cambridge, MA: Harvard University Press, 2007.

Science for All: The Popularization of Science in Early Twentieth-Century Britain. University of Chicago Press, 2009.

'Popular Science Magazines in Interwar Britain: Authors and Readerships.' *Science in Context*, 26 (2013): 437–57.

'Discovering Science from an Armchair: Popular Science in British Magazines of the Interwar Years.' *Annals of Science*, 73 (2016): 89–107.

Boyer, Paul. *By the Bomb's Early Light: American Thought and Culture at the Dawn of the Atomic Age.* New York: Pantheon Books, 1985.

Bradbury, Ray. *The Silver Locusts.* London: Corgi, 1956. [1950].

Branford, Victor. *Science and Sanctity: A Study in the Scientific Approach to Unity.* London: Le Play House and Williams & Norgate, 1923.

Brandau, Daniel. 'Cultivating the Cosmos: Spaceflight Thought in Imperial Germany.' *History and Technology*, 28 (2012): 225–54.

Braun, Wernher von, Fred L. Whipple and Willy Ley. *Man on the Moon.* London: Sidgwick & Jackson, 1953.

Brittain, Vera. *Halcyon: Or the Future of Monogamy.* London: Kegan Paul, 1929.

Broks, Peter. *Media Science before the Great War.* London: Macmillan, 1996.

Understanding Popular Science. Maidenhead: Open University Press, 2006.

Brosnan, John. *Future Tense: The Cinema and Science Fiction.* London: Macdonald and Jane's, 1978.

Brown, Harrison. *The Challenge of Man's Future: An Inquiry Concerning the Condition of Man During the Years That Lie Ahead.* London: Secker and Warburg, 1954.

Bryson, Bill. *One Summer: America 1927.* London: Black Swan, 2013.

Bud, Robert. *The Uses of Life: A History of Biotechnology.* Cambridge University Press, 1993.

Burkett, Jodi. 'The Campaign for Nuclear Disarmament and Changing Attitudes towards the Environment in the Nuclear Age.' *British Journal for the History of Science*, 45 (2012): 625–39.

Burney, Charles Denniston. *The World, the Air and the Future.* London: Alfred A. Knopf, 1929.

Burnham, John C. *How Superstition Won and Science Lost: Popularizing Science and Health in the United States.* New Brunswick, NJ: Rutgers University Press, 1987.

Bush, Donald J. *The Streamlined Decade.* New York: George Braziller, 1975.

Byrd, Richard E. *Skyward.* Ed. William R. Anderson. Chicago: The Lakeside Press, 1981 [1928].

Bywater, Hector C. *The Great Pacific War: A History of the American–Japanese Campaign of 1931–33.* London: Constable, 1925.

Sea-Power in the Pacific: A Study of the American–Japanese Naval Problem. New edn. London: Constable, 1934.

Calder, Peter Ritchie. *The Birth of the Future.* Foreword by F. Gowland Hopkins. London: Arthur Barker, 1934a.

The Conquest of Suffering. Introd. J. B. S. Haldane. London: Methuen, 1934b.

The Life Savers. London: Hutchinson, 1962.

Living with the Atom. University of Chicago Press, 1962.
Hurtling towards 2000 A.D. Derby: Derbyshire Education Authority/University of Nottingham, 1969.
Calder, Nigel (ed.), *The World in 1984: The Complete New Scientist Series*. London: Pelican, 1965, 2 vols.
(ed.), *Technopolis: Social Control of the Uses of Science*. London: Macgibbon & Kee, 1969.
Campbell, John. *Rutherford: Scientist Supreme*. Christchurch, NZ: AAS Publications, 1999.
Campbell, John W. *The Atomic Story*. New York: Henry Holt, 1947.
Campos, Louis A. *Radium and the Secret of Life*. University of Chicago Press, 2015.
Campion, Sarah. *Thirty Million Gas Masks*. London: Peter Davies, 1937.
Cantril, Hadley. *The Invasion from Mars: A Study in the Psychology of Panic: With the Complete Script of the Famous Orson Welles Broadcast*. Princeton University Press, 1940.
Čapek, Karel. *Rossum's Universal Robots (R.U.R.)*. Trans. David Short. London: Modern Voices, 2011 [1923].
Carey, David. *How It Works: The Rocket*. Loughborough: Wills & Hepworth, 1969.
Carlson, W. Bernard. *Tesla: Inventor of the Electrical Age*. Princeton University Press, 2013.
Carpenter, Charles A. *Dramatists and the Bomb: American and British Playwrights Confront the Nuclear Age, 1945–1964*. Westport, CT: Greenwood Press, 1999.
Carr-Saunders, A. M. *Population*. London: Oxford University Press/Humphrey Milford, 1925.
Carrell, Alexis. *Man the Unknown*. New York: Harper Brothers, 1935.
Carrington, Charles. *Rudyard Kipling: His Life and Work*. Harmondsworth: Penguin, 1976.
Carson, Rachel. *Silent Spring*. Boston, MA: Houghton Mifflin, 1962.
Carter, Paul A. *The Twenties in America*. London: Routledge, 1968.
The Creation of Tomorrow: Fifty Years of Magazine Science Fiction. New York: Columbia University Press, 1977.
Cassedy, Steven. *Connected: How Trains, Genes, Pineapples, Piano Keys and a Few Disasters Transformed Americans and the Dawn of the Twentieth Century*. Stanford University Press, 2014.
Cathcart, Brian. *The Fly in the Cathedral: How a Small Group of Cambridge Scientists Won the Race to Split the Atom*. London: Viking, 2004.
Ceadel, Martin. 'Popular Fiction and the Next War, 1918–39' in Frank Gloversmith (ed.), *Class, Culture and Social Change: A New Look at the 1930s*. Sussex: Harvester/New Jersey: Humanities Press, 1980, pp. 161–84.
Ceruzzi, Paul. 'An Unforeseen Revolution: Computers and Expectations, 1935–1985' in Joseph J. Corn (ed.), *Imagining Tomorrow*. Cambridge, MA: MIT Press, 1986, pp. 188–201.
Chadwick, Philip George. *The Death Guard*. London: Hutchison, 1939.
Chapman, James. *British Comics: A Cultural History*. London: Reaktion Books, 2011.
Charles, Enid. *The Twilight of Parenthood: A Biological Study of the Decline of Population Growth*. London: Watts, 1934.
Charlton, L. E. O. *War from the Air: Past, Present, Future*. London: Nelson, 1935.

Chase, Stuart. *Men and Machines*. London: Jonathan Cape, 1929.

Cheng, John. *Astounding Wonder: Imagining Science and Science Fiction in Interwar America*. Philadelphia, PA: University of Pennsylvania Press, 2012.

Chilton, Charles. *Journey into Space*. London: Herbert Jenkins, 1954.

Mission dans l'éspace. Trans. Luc Joye and J.-G. Vandel. Paris: Editions Fleuve Noir, 1959.

Christofleau, André. *Les dernières nouveautés de la science et de l'industrie*. Paris: Hachette, 1925.

Churchill, Winston Spencer. *Thoughts and Adventures*. London: Leo Cooper, 1990 [1932].

Clark, Constance Areson. *God – or Gorilla: Images of Evolution in the Jazz Age*. Baltimore, MD: Johns Hopkins University Press, 2008.

Clark, Ronald W. *JBS: The Life and Work of J. B. S. Haldane*. London: Hodder & Stoughton, 1968.

The Huxleys. London: Heinemann, 1968.

Clarke, Arthur C. *The Exploration of Space*. London: Temple Press, 1951.

The City and the Stars. London: Frederick Muller, 1956.

Profiles of the Future: An Enquiry into the Limits of the Possible. London: Gollancz, 1962.

Prelude to Space. Reprinted London: New English Library, 1977 [1951].

Astounding Days: A Science Fictional Autobiography. London: Gollancz, 1989.

Greetings, Carbon-Based Bipeds!: A Vision of the Twentieth Century As It Happened. London: Harper Collins, 1999.

Clarke, I. F. *The Pattern of Expectation, 1644–2001*. London: Jonathan Cape, 1979.

Voices Prophesying War; Future Wars, 1763–3749. 2nd impression. Oxford University Press, 1992.

Cleator, Philip Ellaby. *Rockets through Space: The Dawn of Inter-Planetary Travel*. London: Allen & Unwin, 1936.

Into Space. London: Allen & Unwin, 1953.

Cockerell, Christopher S. 'The Prospects for Hover Transport' in Nigel Calder (ed.), *The World in 1984*. London: Penguin, 1965, vol. 1, pp. 179–82.

Cogdell, Christina. *Eugenic Design: Streamlining America in the 1930s*. Philadelphia, PA: University of Pennsylvania Press, 2004.

Collard, William. *Proposed London and Paris Railway: London and Paris in 2 Hours 45 Minutes*. Westminster: P. S. King, 1928.

Commoner, Barry. *Science and Survival*. London: Gollancz, 1966.

Conekin, Becky E. *'The Autobiography of a Nation': The 1951 Festival of Britain*. Manchester University Press, 2003.

Connington, J. T. [Alfred Walter Stewart]. *Nordenholt's Million*. London: Constable, 1923.

Corn, Joseph J. *The Winged Gospel: America's Romance with Aviation, 1900–1950*. New York: Oxford University Press, 1983.

(ed.). *Imagining Tomorrow: History, Technology and the American Future*. Cambridge, MA: MIT Press, 1986.

Corn, Joseph J. and Brian Horrigan. *Yesterday's Tomorrows: Past Visions of the American Future*. Baltimore, MD: Johns Hopkins University Press, 1984.

Corelli, Marie. *The Young Diana: An Experiment of the Future*. London: Hutchinson, 1918.

Cowan, Ruth Schwartz. *More Work for Mother: The Ironies of Household Technology from the Open Hearth to the Microwave*. New York: Basic Books, 1983.

Craig, Thurlow. *Plague over London*. London: Hutchinson, 1939.

Cram, Ralph Adams. *Towards the Great Peace*. London: George Harrap, n.d.

Crampsey, Bob. *The Empire Exhibition of 1938: The Last Durbar*. Edinburgh: Mainstream Publishing, 1988.

Cranmer-Byng, L. *Tomorrow's Star: An Essay on the Shattering and Remoulding of the World*. London: Golden Cockerell Press, 1938.

Crompton, Alastair. *The Man Who Drew Tomorrow*. London: Who Dares Publishing, 1985.

Crossley, Robert. *Olaf Stapledon: Speaking for the Future*. Foreword by Brian W. Aldiss. Liverpool University Press, 1994.

Crowther, J. G. 'Breaking up the Atom.' *Nineteenth Century*, 112 (July 1932): 81–94.

Currell, Susan. *American Culture in the 1920s*. Edinburgh University Press, 2009.

and Christina Cogdell (eds). *Popular Eugenics: National Efficiency and American Mass Culture in the 1930s*. Athens, OH: Ohio State University Press, 2006.

Curry, Helen Anne. *Evolution Made to Order: Technological Innovation in Twentieth-Century America*. University of Chicago Press, 2016.

Dalton, Moray. *The Black Death*. London: Sampson Low, Marston, 1934.

Danchev, Alex. *Alchemist of War: The Life of Basil Liddell Hart*. London: Weidenfeld & Nicolson, 1998.

Darwin, Charles Galton. *The Next Million Years*. London: Rupert Hart-Davis, 1952.

The Problems of World Population. Cambridge University Press, 1958.

Davis, Frederick Rowe. *Banned: A History of Pesticides and the Science of Toxicology*. New Haven, CT: Yale University Press, 2014.

Davy, M. J. Bernard. *Air Power and Civilization*. London: Allen & Unwin, 1941.

De Forest, Lee. 'Communication' in Charles A. Beard (ed.), *Toward Civilization*. London: Longmans Green, 1930, pp. 127–36.

De Gaulle, Charles. *The Army of the Future*. London: Hutchinson, 1940.

Del Sesto, Stephen L. 'Wasn't the Future of Nuclear Energy Wonderful?' in Joseph J. Corn (ed.), *Imagining Tomorrow: History, Technology and the American Future*. Cambridge, MA: MIT Press, 1986, pp. 58–76.

Desmond, Shaw. *Ragnarok*. London: Duckworth, 1926.

Devine, Robert A. *Blowing in the Wind: The Nuclear Test Ban Debate, 1954–1960*. New York: Oxford University Press, 1978.

De Vries, Hugo. *Species and Varieties: Their Origin by Mutation*. Chicago: Open Court, 3rd edn. 1912 [1904].

Dewey, Anne Perkins. *Robert Goddard: Space Pioneer*. Boston, MA: Little Brown, 1962.

Dick, Stephen J. *The Biological Universe: The Twentieth-Century Extraterrestrial Life Debate and the Limits of Science*. Cambridge University Press, 1996.

Dietz, David. *Atomic Energy Now and Tomorrow*. London: Westinghouse, 1946.

Douglas, Susan J. 'Amateur Operators and American Broadcasting: Shaping the Future of Radio' in Joseph J. Corn (ed.), *Imagining Tomorrow; History, Technology and the American Future*. Cambridge, MA: MIT Press, 1986, pp. 35–57.

Inventing American Broadcasting, 1899–1922. Baltimore, MD: Johns Hopkins University Press, 1987.

Listening In: Radio and the American Imagination. Minneapolis, MT: University of Minnesota Press, 2004.

Douhet, Giulio. *The Command of the Air.* Trans. Dino Ferrari. London: Faber & Faber, 1933.

Dreaper, W. P. *The Future of Civilisation and Social Science: A Study Involving the Principles of Scientific Meliorism.* London: E. T. Heron, 1937.

Dronamraju, Krishna R. (ed.). *Haldane's Daedalus Revisited.* Oxford University Press, 1995.

Duhamel, Georges. *Scènes de la vie futur.* Paris: Mercure de France, 1939 [1931].

America the Menace: Scenes from the Life of the Future. Trans. Charles Miner Thompson. Boston, MA: Houghton Mifflin, 1931.

Dunnett, Oliver. 'Patrick Moore, Arthur C. Clarke and British "Outer Space" in the Mid-Twentieth Century.' *Cultural Geographies,* 19 (2012): 505–22.

Dunsany, Lord. *Lord Adrian.* Waltham Saint Lawrence, Berks.: Golden Cockerel Press, n.d. [1933].

East, Edward M. *Mankind at the Crossroads.* New York: Scribners, 1923.

Edgell, G. H. *The American Architecture of To-Day.* New York: Scribners, 1928.

Edgerton, David. *England and the Aeroplane: An Essay on a Militant and Technological Nation.* Basingstoke: Macmillan Academic, 1991.

Warfare State: Britain, 1920–1970. Cambridge University Press, 2006.

Edmonds, Harry. *The Professor's Last Experiment.* London: Rich & Cowan, 1935.

Ehrlich, Paul. *The Population Bomb.* London: Ballantine/Friends of the Earth, 1971 [1968].

Eldridge, David. *American Culture in the 1930s.* Edinburgh University Press, 2008.

Endersby, Jim. 'Mutant Utopias: Evening Primroses and Imagined Futures in Early Twentieth-Century America.' *Isis,* 104 (2013): 471–503.

Evans, I. O. *The World of To-Morrow: A Junior Book of Forecasts.* London: Denis Archer, 1933.

Evans, Ronald. *Project Greenglow and the Search for Gravity Control.* London: Matador, 2015.

Farmelo, Graham. *Churchill's Bomb: A Hidden History of Science, War and Politics.* London: Faber & Faber, 2013.

Farry, James and David A. Kirby. 'The Universe Will Be Televised: Space, Science, Satellites and British Television Production, 1946–1969.' *History and Technology,* 28 (2012): 311–33.

Ferris, Hugh. *The Metropolis of Tomorrow.* New York: Ives Washburn, 1929, reprinted Princeton Architectural Press, 1986.

Fielder, Leslie A. *Olaf Stapledon: A Man Divided.* Oxford University Press, 1983.

Fieldler, Jean and Jim Mele. *Isaac Asimov.* New York: Frederick Unger, 1982.

Fleming, James Rodger. *Fixing the Sky: The Checkered History of Weather and Climate Control.* New York: Columbia University Press, 2010.

Foertsch, Jacqueline. *American Culture in the 1940s.* Edinburgh University Press. 2008.

Forbes-Sempill, William (The Master of Sempill). *The Air and the Plain Man.* London: Elkin Mathews & Marrot, 1931.

Ford, Henry. *Today and Tomorrow.* London: Heinemann, 1926.

Forgan, Sophie. 'Festivals of Science and the Two Cultures: Science, Design and Display at the Festival of Britain, 1951.' *British Journal for the History of Science,* 31 (1998): 217–40.

'Atoms in Wonderland.' *History and Technology*, 19 (2003): 177–96.

Forster, E. M. *Collected Short Stories*. London: Penguin Classics, 2002.

Foster, Jonathan. *The Death Ray: The Secret Life of Harry Grindell Matthews*. London: Invention Publishing, 2008.

Foster, Norman. *Dymaxion Car: Buckminster Fuller*. London and Madrid: Ivory Press, 2011.

Fournier d'Albe, E. E. *Hephaestus, or, The Soul of the Machine*. London: Kegan Paul, Trench, Trubner, 1925.

Quo Vadimus? Some Glimpses of the Future. London: Kegan Paul, 1925.

Frankl, Paul T. *Form and Re-Form: A Practical Handbook of Modern Interiors*. New York: Harper, 1930.

Franklin, H. Bruce. *Robert A. Heinlein: America as Science Fiction*. New York: Oxford University Press, 1980.

'America as Science Fiction: 1939' in George E. Slusser, Eric S. Rabkin and Robert Scholes (eds), *Coordinates: Placing Science Fiction and Fantasy*. Carbondale and Edwardsville: Southern Illinois University Press, 1983, pp. 107–23.

War Stars: The Superweapon and the American Imagination. Revised and extended edn. Amherst: University of Massachusetts Press, 2008.

Frayling, Christopher. *Mad, Bad and Dangerous? The Scientist and the Cinema*. London: Routledge, 2005.

Freud, Sigmund. *The Future of an Illusion*. Trans. W. D. Rollston-Scott. London: Hogarth Press, 1962.

Fuller, J. F. C. *The Reformation of War*. London: Hutchinson, 1923.

Atlantis: America and the Future. London: Kegan Paul, n.d.

Pegasus: Problems of Transportation. London: Kegan Paul, n.d.

On Future Warfare. London: Sifton Praed, 1928.

Furnas, C. C. *The Next Hundred Years: The Unfinished Business of Science*. London: Cassell, 1936.

Gabor, Dennis. *Inventing the Future*. London: Secker & Warburg, 1963.

Gamow, George. *Atomic Energy in Cosmic and Human Life*. Cambridge University Press, 1947.

Garrett, Garet. *Ouroboros: Or the Mechanical Extension of Mankind*. London: Kegan Paul, n.d.

Gelenter, David. *1939: The Lost World of the Fair*. New York: Free Press, 1995.

Geddes, Norman Bel. *Horizons*. London: John Lane, the Bodley Head, 1932.

Geddes, Patrick. *Cities in Evolution: An Introduction to the Town Planning Movement and to the Study of Cities*. London: Williams & Norgate, 1915.

Gernsback, Hugo. 'Fifty Years from Now.' *Tit-Bits* (14 March 1925): 59.

Geppert, Alexander C. T. 'Space *Personae*: Cosmopolitan Networks of Peripheral Knowledge, 1927–1957.' *Journal of Modern European History*, 6 (2008): 262–85.

(ed.). *Imagining Outer Space: European Astroculture in the Twentieth Century*. Basingstoke: Palgrave Macmillan, 2012.

Gibbs, Philip. *The Day after Tomorrow: What is Going to Happen to the World?* London: Hutchinson, 1928.

Gibbs-Smith, Charles Harvard. *Aviation: An Historical Survey from Its Origins to the End of World War II*. London: HMSO, 1970.

Gifford, James. *Robert A. Heinlein: A Reader's Companion*. Sacramento: Nitrosyncretic Press, 2000.

Glancey, Jonathan. *Concorde: The Rise and Fall of the Supersonic Airliner*. London: Atlantic Books, 2015.

Gloag, John. *Artifex: Or the Future of Craftsmanship*. London: Kegan Paul, n.d.

Winter's Youth. London: Allen & Unwin. 1934.

Goddard, Esther C. and G. Edward Pendray (eds). *The Papers of Robert H. Goddard*. New York: McGraw Hill, 1970, 3 vols.

Godfrey, Hollis. *The Man Who Ended War*. London: Ward Lock, 1910.

Godwin, George. *Empty Victory*. London: John Lang, 1932.

Goldsmith, Maurice. *Sage: A Life of J. D. Bernal*. London: Hutchinson, 1980.

Gooday, Graeme. *Domesticating Electricity: Technology, Uncertainty and Gender, 1884–1914*. London: Pickering & Chatto, 2008.

Graham, Loren. *Lysenko's Ghost: Epigenetics in Russia*. Cambridge, MA: Harvard University Press, 2016.

Grahame-White, Claude and Harry Harper. *The Aeroplane*. London: T. C. and E. C. Jack, 1914.

Air Power: Naval, Military Commercial. London: Chapman & Hall, 1917.

Graves, Robert and Alan Hodge. *The Long Week-End: A Social History of Great Britain, 1918–1939*. London: Faber & Faber, 1940.

Green, Roger Lancelyn and Walter Hooper. *C. S. Lewis: A Biography*. London: Collins, 1974.

Greenhalgh, Paul. *Ephemeral Vistas: The* Expositions Universelles, *Great Exhibitions and World Fairs, 1851–1939*. Manchester University Press, 1988.

Gregory, Jane and Steve Miller. *Science in Public: Communication, Culture and Credibility*. Cambridge, MA: Perseus Publishing, 2000.

Gregory, Owen. *Meccania: The Super State*. London: Methuen, 1918.

Guggenheim, Harry F. *The Seven Skies*. New York: Putnam's Sons, 1930.

Haapamaki, Michele. *The Coming of the Aerial War: Culture and the Fear of Airborne Attack in Inter-War Britain*. London: I. B. Tauris, 2014.

Haldane, J. B. S. *Daedalus: Or Science and the Future*. London: Kegan Paul, 1924.

Possible Worlds: And Other Essays. London: Chatto & Windus, 1930.

'Auld Hornie, F.R.S.' *Modern Quarterly*, n.s. (1946): 32.

'Biological Possibilities for the Human Species in the Next Ten Thousand Years' in Gordon Wolstenholme (ed.), *Man and his Future: A CIBA Foundation Volume*. London: J. and A. Churchill, 1963, pp. 337–61.

Hall, Diana Long. 'The Critic and the Advocate: Contrasting British Views on the State of Endocrinology in the Early 1920s.' *Journal of the History of Biology*, 9 (1976): 269–85.

Halsbury, the Earl of [Hardinge Goulburn Gifford]. *1944*. London: Thornton Butterworth, 1926.

Hamilton, David. *The Monkey Gland Affair*. London: Chatto & Windus, 1986.

Hamilton-Patterson, James. *Empire of the Clouds: When Britain's Aircraft Ruled the World*. London: Faber & Faber, 2010.

Hampton, Kirk and Carol MacKay. 'The Internet and the Analogical Myths of Science Fiction' in Gary Westfahl, Wong Kin Yuen and Amy Kit-Sze Chan (eds), *Science Fiction and the Prediction of the Future*. Jefferson, NC: McFarland, 2011, pp. 41–51.

Hanks, David A. and Anna Hoy. *American Streamlined Design: The World of Tomorrow*. Paris: Flammarion, 2005.

Hansen, Bert. *Picturing Medical Progress from Pasteur to Polio: A History of Mass Media Images and Popular Attitudes in America*. New Brunswick, NJ: Rutgers University Press, 2009.

Harper, Harry. *Dawn of the Space Age*. London: Sampson Low, Marston & Co, 1946.

Harris, J. P. *Men, Ideas and Tanks: British Military Thought and Armoured Forces, 1903–1939*. Manchester University Press, 1995.

Harrison, Helen A. (ed.), *Dawn of a New Day: The New York World's Fair, 1939/40*. New York: Queen's Museum/New York University Press, 1980.

Hart, Basil Liddell. *Paris: Or the Future of War*. London: Kegan Paul, 1925.

When Britain Goes to War: Adaptability and Mobility. London: Faber & Faber, 1932.

Hartley, Olga and Mrs C. F. Leyel. *Lucullus: Or the Food of the Future*. London: Kegan Paul, 1927.

Hatfield, H. Stafford. *Automaton: Or the Future of the Mechanical Man*. London: Kegan Paul, 1928.

The Inventor and His World. London: Kegan Paul, 1933.

Haynes, Rosslyn D. *H. G. Wells: Discoverer of the Future*. London: Macmillan, 1980.

From Faust to Strangelove: Representations of the Scientist in Western Literature. Baltimore, MD: Johns Hopkins University Press, 1994.

Heinlein, Robert A. *The Man Who Sold the Moon*. London: Sidgwick & Jackson, 1953.

The Moon is a Harsh Mistress. London: Hodder & Stoughton, 1967 [1966].

Starman Jones. London: New English Library, 1976 [1953].

Hewison, Robert. *Culture and Consensus: England, Art and Politics since 1940*. London: Methuen, 1995.

Hillegas, Mark R. *The Future as Nightmare: H. G. Wells and the Anti-Utopians*. New York: Oxford University Press, 1967.

Hobsbawm, Eric. 'C (for Crisis).' *London Review of Books*, 31 (6 August 2009): 12–13.

Hodgson, John. *The Time-Journey of Dr Barton: An Engineering and Sociological Forecast Based on Present Possibilities*. Eggington, Berks.: John Hodgson, 1929.

Holman, Brett. 'The Shadow of the Airliner: Commercial Bombers and the Rhetorical Destruction of Britain, 1917–35.' *Twentieth-Century British History*, 24 (2013): 495–517.

The Next War in the Air: Britain's Fear of the Bomber, 1908–1941. Farnham, Surrey: Ashgate, 2014.

Horniman, Roy. *How to Make the Railways Pay for the War; or the Transport Problem Solved*. London: Routledge, 1919.

Horrigan, Brian. 'The Home of Tomorrow' in Joseph J. Corn (ed.), *Imagining Tomorrow: History, Technology and the American Future*. Cambridge, MA: MIT Press, 1986, pp. 137–63.

Howard, Ebenezer. *Garden Cities of To-Morrow*. Introduced by Lewis Mumford. London: Faber & Faber, 1946 [1902].

Huizinga, J. *In the Shadow of Tomorrow: A Diagnosis of the Spiritual Distemper of Our Time*. London: Heinemann, 1936.

Hunter, I. Q. *British Science Fiction Cinema*. London: Routledge, 1999.

Huntington, John. *The Logic of Fantasy: H. G. Wells and Science Fiction*. New York: Columbia University Press, 1982.

Hutchings, Peter. 'We're All Martians Now: British Invasion Fantasies of the 1950s and 1960s' in I. Q. Hunter (ed.), *British Science Fiction Cinema*. London: Routledge, 1999, pp. 33–47.

Hussingtree, Martin. *Konyetz*. London: Hodder & Stoughton, 1924.

Huxley, Aldous. *Chrome Yellow*. Harmondsworth: Penguin, 1955 [1921].

Brave New World Revisited. London: Chatto & Windus, 1959.

Brave New World: A Novel. Harmondsworth: Penguin, 1962 [1932].

After Many a Summer. London: Vintage, 2015 [1939].

Huxley, Julian S. *Memories*. London and New York: Harper & Row, 1970.

Inge, Willam Ralph. 'The Future of the Human Race'. *Proceedings of the Royal Institution of Great Britain*, 26 (1929–31): 494–515.

Ingram, Kenneth. *The Coming Civilization: Will It Be Capitalist? Will It Be Materialist?* London: Allen & Unwin, 1935.

Inman, Mason. *The Oracle of Oil: A Maverick Geologist's Quest for a Sustainable Future*. New York: Norton, 2016.

Jackson, Robert. *Airships in Peace and War*. London: Cassell, 1971.

James, Edward and Frank Mendlesohn (eds), *The Cambridge Companion to Science Fiction*. Cambridge University Press, 2003.

Jay, Kenneth. *Calder Hall: The Story of Britain's First Atomic Power Station*. London: Methuen, 1956.

Jennings, H. S. *Prometheus: Or Biology and the Advancement of Man*. London: Kegan Paul, n.d.

Jeremiah, David. *Architecture and Design for the Family in Britain, 1900–70*. Manchester University Press, 2000.

Joad, C. E. M. (ed.), *Manifesto: Being the Book of the Federation of Progressive Societies and Individuals*. London: Allen & Unwin, 1934.

Johns, W. E. *Kings of Space: A Story of Interplanetary Exploration*. London: Hodder & Stoughton, 1954.

Return to Mars: A Story of Interplanetary Flight. London: Hodder & Stoughton, 1955.

Johnson, Brian. *The Secret War*. London: BBC, 1978.

Jolly, W. P. *Sir Oliver Lodge: Psychical Researcher and Scientist*. London: Constable, 1974.

Jones, T. W. *Hermes: or the Future of Chemistry*. London: Kegan Paul, 1928.

Joravsky, D. *The Lysenko Affair*. Cambridge, MA: Harvard University Press, 1970.

Kaempffert, Waldemar. *Science Today and Tomorrow*. London: Nicholson Watson, 1940.

Science Today and Tomorrow: Second Series. London: Denis Dobson, 1947.

Explorations in Science. London: Gollancz, 1953.

Kahn, Herman. *On Thermonuclear War*. Princeton University Press, 1960.

Kahn, Herman and Anthony J. Wiener. *The Year 2000: A Framework for Speculation on the Next Thirty-Three Years*. New York: Macmillan, 1967.

'The Next Thirty-Three Years: Framework for Speculation.' *Daedalus*, 96 (1967): 705–32.

Kamm, Anthony and Malcolm Baird. *John Logie Baird: A Life*. Edinburgh: National Museum of Scotland, 2002.

Kammerer, Paul. *Rejuvenation and the Prolongation of Human Efficiency: Experiences with the Steinach-Operation on Man and Animals*. London: Methuen, 1924.

Kargon, Robert. *The Rise of Robert Millikan: Portrait of a Life in American Science.* Ithaca, NY: Cornell University Press, 1982.

Kargon, Robert, Karen Fiss, Martin Low and Arthur P. Moellan. *World's Fairs on the Eve of War.* University of Pittsburgh Press, 2015.

Karp, David. *One.* Yardley, PA: Westholme Publishers, 1953.

Kenworthy, J. M. *New Wars: New Weapons.* London: Elkin Mathews & Marrot, 1930.

Kevles, Daniel. *In the Name of Eugenics: Genetics and the Uses of Human Heredity.* New York: Knopf, 1985.

Kilgore, De Witt Douglas. *Astrofuturism: Science, Race, and Visions of Utopia in Space.* Philadelphia, PA: University of Pennsylvania Press, 2003.

Kingsland, Sharon E. 'The Battling Botanist: Daniel Trembly MacDougal, Mutation Theory, and the Rise of Experimental Evolutionary Biology in America, 1900–1912.' *Isis,* 82 (1991): 479–509.

Kipling, Rudyard. *Actions and Reactions.* London: Macmillan, 1909.

A Diversity of Creatures. London: Macmillan, 1917.

Kitchen, Paddy. *A Most Unsettling Person: An Introduction to the Life and Ideas of Patrick Geddes.* London: Victor Gollancz, 1975.

Klerkx, Greg. *Lost in Space: The Fall of NASA and the Dream of a New Space Age.* London: Secker & Warburg, 2004.

Knibbs, George Handley. *The Shadow of the World's Future: Or the Earth's Population Possibilities & the Consequences of the Present Rate of Increase of the Earth's Inhabitants.* London: Ernest Benn, 1928.

Knight, Donald R. and Alan D. Sabey. *The Lion Roars at Wembley: British Empire Exhibition 60th Anniversary.* London: Barnard & Westwood, 1984.

Knight, Nancy. '"The New Light": X Rays and Medical Futurism' in Joseph J. Corn (ed.), *Imagining Tomorrow: History, Technology and the American Future.* Cambridge, MA: MIT Press, 1986, pp. 10–34.

Koestler, Arthur. *The Case of the Midwife Toad.* London: Hutchinson, 1971.

Kohl, Leonard J. 'Flash Gordon Conquers the Great Depression and World War Too!' in David J. Hogan (ed.), *Science Fiction America: Essays in SF Cinema.* Jefferson, NC: MacFarland, 2006, pp. 40–56.

Köhlstedt, Folke T. 'Utopia Realized: The World's Fairs of the 1930s' in Joseph J. Corn (ed.), *Imagining Tomorrow.* Cambridge, MA: MIT Press, 1986, pp. 97–118.

Kohn, Robert D. 'Social Ideals in a World's Fair' in Warren Susman (ed.), *Culture and Commitment: 1929–1945.* New York: George Braziler, 1973, pp. 297–300 [1939].

Kramers, H. A. and Helga Holst. *The Atom and the Bohr Theory of Its Structure.* With a foreword by Sir Ernest Rutherford. London: Gylendal, 1923.

Krementsov, Nikolai. *Revolutionary Experiments: The Quest for Immortality in Bolshevik Science and Fiction.* New York: Oxford University Press, 2014.

La Follette, Marcel. *Making Science Our Own: Public Images of Science, 1910–1955.* University of Chicago Press, 1990.

Science on the Air: Popularizers and Personalities on Radio and Early Television. University of Chicago Press, 2008.

Science on American Television: A History. University of Chicago Press, 2013.

Lanchester, F. W. *Aircraft in Warfare: The Dawn of the Fourth Arm.* London: Constable, 1916.

Langdon-Davies, John. *A Short History of the Future.* London: Routledge, 1936.

Lapp, Ralph E. *The New Priesthood: The Scientific Elite and the Uses of Power.* New York: Harper & Row, 1965.

Lasser, David. *The Conquest of Space.* Foreword by Dr H. H. Sheldon. London: Hurst & Blackett, 1932.

Laurence, William L. *Dawn over Zero: The Story of the Atomic Bomb.* London: Museum Press, 1947.

Le Bon, Gustav. *The World in Revolt: A Psychological Study of our Times.* Trans. Bernard Miall. London: T. Fisher Unwin, 1921.

Le Corbusier. *The City of To-Morrow and Its Planning.* Trans. Frederick Etchells. London: Architectural Press, 1971 [1924, 1929].

Lee, Gerald Stanley. *Crowds.* London: Curtis & Brown, 1913.

Lefebure, Victor. *Scientific Disarmament.* London: Mundanus (Victor Gollancz), 1931.

Leonard, Jonathan Norton. *Tools of Tomorrow.* London: Routledge, 1935.

Le Queux, William. *The Invasion of 1910: With a Full Account of the Siege of London.* London: E. Nash, 1906.

The Terror of the Air. London: Lloyds, 1920.

Lessing, Lawrence (ed.), *DNA: At the Core of Life Itself.* New York: Macmillan, 1966.

Lewis, C. S. *The Cosmic Trilogy: Out of the Silent Planet; Perelandra; That Hideous Strength.* London: Bodley Head, 1989.

Of Other Worlds: Essays and Stories. Ed. Walter Hooper. London: Geoffrey Bles, 1966.

Ley, Willy. *Rockets and Space Travel: The Future of Flight beyond the Stratosphere.* London: Chapman & Hall, 1948.

The Conquest of Space. Paintings by Charles Bonestrell. London: Sidgwick & Jackson, 1950.

Rockets, Missiles, and Space Travel. London: Chapman & Hall, 1957.

Ley, Willy and Wernher von Braun. *The Exploration of Mars.* London: Sidgwick & Jackson, 1956.

Link, Eric Carl and Gerry Canavan (eds), *The Cambridge Companion to American Science Fiction.* Cambridge University Press, 2015.

Lockhart-Mummery, J. Percy. *After Us: Or the World As It Might Be.* London: L. Stanley Paul, 1936.

London, Jack. *The Iron Heel.* Preface by Anatole France. 12th edn. London: Mills & Boon, 1932.

Low, A. M. *Wireless Possibilities.* London: Kegan Paul, 1924.

The Future. London: Routledge, 1925.

Our Wonderful World of Tomorrow: A Scientific Forecast of the Men, Women and the World of the Future. London: Ward Lock, 1934.

Adrift in the Stratosphere. London: Blackie, 1937.

Mars Breaks Through: Or The Great Murchison Mystery. London: Herbert Joseph, n. d. [1938].

Electronics Everywhere. London: Museum Press, 1951.

Luckett, Perry D. *Charles A. Lindbergh: A Bio-Bibliography.* New York: Greenwood Press, 1986.

Ludovici, A. M. *Lysistrata: Woman's Future and Future Woman.* London: Kegan Paul, n.d.

Macaulay, Rose. *What Not: A Prophetic Comedy.* London: Constable, 1918.

Macaulay, William R. '"Crafting the Future": Envisioning Space Exploration in Post-War Britain.' *History and Technology*, 28 (2012): 281–309.

MacFie, Ronald Campbell. *Metanthropos: Or the Body of the Future*. London: Kegan Paul, 1928.

Mackworth-Praed, Ben. *Aviation: The Pioneer Years*. London: Studio Editions, 1990.

Macmillan, Harold. *Winds of Change: 1914–1939*. London: Macmillan, 1966.

Mahaffey, James A. *Atomic Accidents: A History of Nuclear Meltdowns and Disasters*. New York: Pegasus Books, 2015.

Malan, Lloyd. *Men, Rockets and Space*. London: Cassell, 1956.

Manduco, Joseph. *The Meccano Magazine, 1916–1983*. London: New Cavendish Books, 1987.

Manning, Laurence. *The Man Who Awoke*. Reprinted London: Sphere Books, 1977 [1933].

Marks, Robert and R. Buckminster Fuller. *The Dymaxion World of Buckminster Fuller*. Garden City, NY: Anchor Press/Doubleday, 1973.

Marsden, Ben and Crosbie Smith. *Engineering Empires: Technology, Science, and Culture, 1760–1911*. London: Palgrave Macmillan, 2005.

Martin, Camille. 'The Railways of Tomorrow' in Nigel Calder (ed.), *The World in 1984*. London: Penguin, 1965, vol. 1, pp. 183–6.

Marvell, Andrew. *Three Men Make a World*. London: Gollancz, 1939.

Marvin, F. S. *The New Vision of Man*. London: Allen & Unwin, 1938.

Maurer, Eva, Julia Richards, Monica Rüthers and Carmen Scheide (eds). *Soviet Space Culture: Cosmic Enthusiasm in Socialist Societies*. London: Palgrave Macmillan, 2011.

Mazower, Mark. *Governing the World: The History of an Idea*. New York: Penguin, 2012.

McAleer, Neil. *Odyssey: The Authorized Biography of Arthur C. Clarke*. London: Gollancz, 1992.

McCray, W. Patrick. *The Visioneers: How a Group of Elite Scientists Pursued Space Colonies, Nanotechnologies and a Limitless Future*. Princeton University Press, 2013.

McCurdy, Howard E. *Space and the American Imagination*. Washington: Smithsonian Institution Press, 1997.

McGrady, Patrick M., Jr. *The Youth Doctor*. London: Arthur Barker, 1968.

McIlraith, Frank and Roy Connolly. *Invasion from the Air: A Prophetic Novel*. London: Grayson & Grayson, 1934.

McLaren, Angus. *Reproduction by Design: Sex, Robots, Trees and Test Tube Babies in Interwar Britain*. University of Chicago Press, 2012.

Mee, Arthur (ed.), *Harmsworth Popular Science*. London: Amalgamated Press, 1911–12. Reissued London: Educational Book Co., 1914, 7 vols.

Meikle, Jeffrey L. 'Plastic, Material of a Thousand Uses' in Joseph. J. Corn (ed.), *Imagining Tomorrow*. Cambridge, MA: MIT Press, 1986, pp. 77–96.

American Plastic: A Cultural History. New Brunswick, NJ: Rutgers University Press, 1995.

Twentieth-Century Limited: Industrial Design in America, 1925–1939. 2nd edn. Philadelphia, PA: Temple University Press, 2001.

Merricks, Linda. *The World Made New: Frederick Soddy, Science, Politics and Environment*. Oxford University Press, 1996.

Merriman, Peter. 'A Power for Good or Evil: Geographies of the M1 in Late Fifties Britain' in David Gilbert, David Matless and Brian Short (eds), *Geographies of British Modernity: Space and Society in the Twentieth Century*. Oxford: Blackwell, 2003, pp. 115–31.

Miller, Walter M., Jr. *A Canticle for Leibowitz*. Reprinted London: Gollancz, 2013 [1960].

Millikan, Robert A. 'Science Lights the Torch' in Charles A. Beard (ed.), *Toward Civilization*. London: Longmans Green, 1930.

Science and the New Civilization. New York: Scribners, 1930.

Mitchell, J. Leslie. *Hanno: Or the Future of Exploration*. London: Kegan Paul, 1928.

Gay Hunter. London: Heinemann, 1934.

Mitchell, William. *Winged Defense: The Development and Possibilities of Modern Air Power – Economic and Military*. New York: Putnam, 1925.

Moffett, Cleveland. *The Conquest of America: A Romance of Disaster and Victory: U.S.A., 1921 A.D.* London: Hodder & Stoughton, 1916.

Moore, Patrick. *Mission to Mars*. London: Burke, 1955.

The Boys' Book of Space. 3rd edn. London: Burke, 1959 [1954].

Morris, Marcus (ed.), *The Best of Eagle*. London: Michael Joseph, 1977.

Morris, Sally and Jan Hallwood. *Living with Eagles: Marcus Morris, Priest and Publisher*. Cambridge: Littleworth Press, 1998.

Muller, H. J. *Out of the Night: A Biologist's View of the Future*. London: Gollancz, 1936.

Mumford, Lewis. *Technics and Civilization*. London: Routledge, 1934.

City Development: Studies in Disintegration and Renewal. London: Secker & Warburg, 1946.

The Culture of Cities. Reprinted London: Routledge/Thoemmes Press, 1997 [1938].

Murray, Nicholas. *Aldous Huxley: An English Intellectual*. London: Little, Brown, 2002.

Myhra, David. *Secret Aircraft Designs of the Third Reich*. Atglen, PA: Schiffer Military/ Aviation History, 1998.

Nelson, Craig. *The Age of Radiance: The Epic Rise and Dramatic Fall of the Atomic Era*. New York: Scribner, 2014.

'Neon' [Marion W. Ackworth]. *The Great Delusion: Study of Aircraft in Peace and War. With a Preface by Arthur Hungerford Pollen*. London: Ernest Benn, 1927.

Neufeld, Michael J. *Von Braun: Dreamer of Space, Engineer of War*. New York: Knopf, 2007.

Newman, Bernard. *Armoured Doves: A Peace Book*. London: Jarrolds, 1936.

Nichols, Beverley. *Cry Havoc!* London: Jonathan Cape, 1933.

Nichols, Peter (ed.). *The Encyclopedia of Science Fiction*. London: Granada, 1974.

Nichols, Robert and Maurice Browne. *Wings over Europe: A Dramatic Extravaganza*. London: Chatto & Windus, 1932.

Nicolson, Harold. *Public Faces*. London: Constable, 1932.

Norton, Roy. *The Vanishing Fleets*. New York: Appleton, 1908.

Novak, Frank G. Jr. *Lewis Mumford and Patrick Geddes: The Correspondence*. London: Routledge, 1955.

Nye, David E. *Electrifying America: Social Meanings of a New Technology, 1880–1940*. Cambridge, MA: MIT Press, 1990.

America's Assembly Line. Cambridge, MA: MIT Press, 2013.

Nye, Mary Jo. 'Gustav Le Bon's Black Light: A Study in Physics and Philosophy in France at the Turn of the Century.' *Historical Studies in the Physical Sciences*, 4 (1974): 163–95.

Ohana, David. *The Futurist Syndrome*. Eastbourne: Sussex Academic Press, 2010.

Olander, Joseph D. and Martin Harry Greenberg (eds), *Arthur C. Clarke*. Edinburgh: Paul Harris, 1977.

(eds), *Isaac Asimov*. Edinburgh: Paul Harris, 1977.

(eds), *Robert A. Heinlein*. Edinburgh: Paul Harris, 1978.

O'Neill, Joseph. *Day of Wrath*. London: Gollancz, 1936.

Oppenheim, Janet. *The Other World: Spiritualism and Psychical Research in England, 1850–1914*. Cambridge University Press, 1985.

Ortolano, Guy. *The Two Cultures Controversy: Science, Literature and Cultural Politics in Postwar Britain*. Cambridge University Press, 2009.

Orwell, George [Eric Bair]. *Nineteen Eighty-Four*. Harmondsworth: Penguin, 1954 [1948].

The Collected Essays, Journalism and Letters of George Orwell, vol. 1: *1920–1940*; vol. 3: *1943–1945*. Ed. Sonia Orwell and Ian Angus. London: Secker & Warburg, 1968. Reprinted London: Penguin, 1970.

Osborn, Fairfield. *Our Plundered Planet*. London: Faber & Faber, 1948.

Overy, Richard. *The Morbid Age: Britain and the Crisis of Civilization*. London: Allen Lane, 2009.

Panchasi, Roxanne. *Future Tense: The Culture of Anticipation in France between the Wars*. Ithaca, NY: Cornell University Press, 2009.

Parrinder, Patrick. *Shadows of the Future: H. G. Wells, Science Fiction and Prophecy*. Liverpool University Press, 1995.

Pauly, Philip J. *Controlling Life: Jacques Loeb and the Engineering Idea in Biology*. New York: Oxford University Press, 1987.

Payne, Lee. *Lighter than Air: An Illustrated History of the Airship*. South Brunswick and New York: A. S. Barnes/London: Thomas Yoselloff, 1997.

Peden, G. C. *Arms, Economics and British Strategy: From Dreadnaughts to Hydrogen Bombs*. Cambridge University Press, 2007.

Pegg, Mark. *Broadcasting and Society, 1918–1939*. London: Croom Helm, 1983.

Pevsner, Nikolaus. *Pioneers of Modern Design: From William Morris to Walter Gropius*. Introd. Richard Weston. Bath: Palazzo, 2011 [1936].

Pohl, Frederick. *The Way the Future Was: A Memoir*. London: Gollancz, 1979.

Pollard, Leslie. *Menace: A Novel of the Near Future*. London: T. Werner Laurie, 1935.

Poole, Robert. 'The Challenge of the Spaceship: Arthur C. Clarke and the History of the Future, 1930–1970.' *History and Technology*, 28 (2012): 255–80.

Priestley, J. B. *The Doomsday Men: An Adventure*. London: Heinemann, 1938.

Pugh, Martin. *We Danced All Night: A Social History of Britain between the Wars*. London: Bodley Head, 2008.

Quester, George H. *Deterrence before Hiroshima: The Airpower Background of Modern Strategy*. New York: Wiley, 1966.

Ramseyer, Edwin. *Airmen over the Suburbs*. Trans. Nora Bickley. London: Gollancz, 1939.

Rathenau, Walther. *In Days to Come*. Trans. Eden Paul and Cedar Paul. London: Allen & Unwin, 1921.

Reade, Winwood. *The Martyrdom of Man*. London: Trubner, 1884 [1872].

Reith, J. C. W. *Broadcast over Britain*. London: Hodder & Stoughton, 1924.

Revelle, Roger. 'A Long View from the Beach' in Nigel Calder (ed.), *The World in 1984*. London: Penguin, 1964, vol. 1, pp. 106–14.

Ritchie-Calder, Lord. *See* Calder, Peter Ritchie.

Ritschel, Daniel. *The Politics of Planning: The Debate on Economic Planning in Britain in the 1930s*. Oxford: Clarendon Press, 1997.

Robin, Theodore. 'The Real Doctor Voronoff.' *Armchair Science*, 1 (1929): 202–3.

Rodwell, Grant. 'Dr. Caleb William Saleeby: The Complete Eugenist.' *History of Education*, 26 (1997): 23–40.

Roger, Noell. *The New Adam*. Trans. P. O. Crowhurst. London: Stanley Paul, 1926.

Rollins, William H. 'Whose Landscape? Technocracy, Fascism and Environmentalism on the National Socialist Autobahn.' *Annals of the Association of American Geographers*, 85 (1995): 494–520.

Ross, Andrew. *Strange Weather: Culture, Science, and Technology in the Age of Limits*. London and New York: Verso, 1991.

Ross, Edward Alsworth. *Standing Room Only?* London: Chapman & Hall, 1928.

Ross, Kristin. *Fast Cars, Clean Bodies: Decolonization and the Reordering of French Culture*. Cambridge, MA: MIT Press, 1995.

Rostand, Jean. *Can Man Be Modified?* Trans. Jonathan Grifffin. London: Secker & Warburg, 1959.

Rousseau, Victor. *The Messiah of the Cylinder*. London: Curtis Brown, 1917.

Ruse, Michael. *Monad to Man: The Concept of Progress in Evolutionary Biology*. Cambridge, MA: Harvard University Press, 1997.

Russell, Bertrand. *Icarus: Or the Future of Science*. London: Kegan Paul, 1924.

The Scientific Outlook. London: Allen & Unwin, 1931.

Russell, Dora. *Hypatia: or Woman and Knowledge*. London: Kegan Paul, 1925.

Russell, Doug. 'Popularization and the Challenge to Science-Centrism in the 1930s' in M. W. McRae (ed.), *The Literature of Science: Perspectives on Popular Science Writing*. Athens, GA: University of Georgia Press, 1993, pp. 37–53.

Rydell, Robert W. *All the World's a Fair: Visions of Empire at American International Expositions, 1876–1916*. University of Chicago Press, 1984.

World of Fairs: The Century-of-Progress Expositions. University of Chicago Press, 1993.

and Laura Burd Schiavo. *Designing Tomorrow: America's World Fairs of the 1930s*. New Haven, CT: Yale University Press, 2010.

Sage, Daniel. 'Framing Space: A Popular Geopolitics of American Manifest Destiny in Outer Space.' *Geopolitics*, 13 (2008): 27–53.

Saleeby, Caleb Williams. *The Conquest of Cancer: A Plan of Campaign: Being an Account of the Principles and Practice hitherto of the Treatment of Malignant Growths by Specific or Cancrotoxic Ferments*. London: Chapman & Hall, 1907.

Sampson, Anthony. *Empires of the Sky: The Politics, Contests and Cartels of World Airlines*. London: Hodder & Stoughton, 1984.

Samuel, Viscount H. L. *An Unknown Land*. London: Allen & Unwin, 1942.

Santomasso, Eugene A. 'The Design of Reason: Architecture and Planning at the 1939/40 New York World's Fair' in Helen A. Harrison (ed.), *Dawn of a New Day*. New York: The Queen's Museum/New York University Press, 1980, pp. 29–40.

Sartre, Pierre. 'Travelling by Air in 1984' in Nigel Calder (ed.), *The World in 1984*. London: Penguin, 1969, vol. 1, pp. 169–74.

Sax, Karl. *Standing Room Only: The Challenge of Over-Population.* Boston, MA: Beacon Press, 1955.

Schiller, F. C. S. *Tantalus: Or the Future of Man.* London: Kegan Paul, 1924.

Schirmacher, Arne. 'From *Kosmos* to *Koralle*: The Culture of Science Reading in Imperial and Weimar Germany' in Catherine Carson, Alexei Kojenikov and Helmuth Trischler (eds), *Quantum Mechanics and Weimar Culture: Revisiting the Forman Thesis.* London: World Scientific, 2010, pp. 43–52.

Schmidt, Peter. *The Conquest of Old Age: Methods to Effect Rejuvenation and to Increase Functional Activity.* London: Routledge, 1931.

Scott, David Meerman and Richard Jurek, *Marketing the Moon.* Cambridge, MA: MIT Press, 2014.

Sears, Paul B. *Deserts on the March.* Norman, OK: University of Oklahoma Press, 1947 [1935].

Segal, Howard P. *Technological Utopianism in American Culture.* University of Chicago Press, 1985.

Sengoopta, Chanak. *The Most Secret Quintessence of Life: Sex, Glands, and Hormones, 1850–1950.* University of Chicago Press, 2006.

Shand, James D. 'The Reichsautobahn: Symbol for the Third Reich.' *Journal of Contemporary History,* 19 (1984): 189–200.

Sheller, Mimi. *Aluminum Dreams: The Making of Light Modernity.* Cambridge, MA: MIT Press, 2014.

Shute, Nevil [Nevil Shute Norway]. *What Happened to the Corbetts.* Reprinted London: Vintage, 2009 [1939].

Slide Rule: The Autobiography of an Engineer. Reprinted London: Vintage, 2009 [1954].

On the Beach. London: Mandarin, 1990 [1957].

Siddiqi, Asif A. *The Red Rockets' Glare: Spaceflight and the Soviet Imagination, 1857–1957.* Cambridge University Press, 2010.

Sims, Phillip E. *Adventurous Empires: The Story of the Empire Flying Boats.* Barnsley: Pen & Sword Books, 2013.

Sinclair, J. A. *Airships in Peace and War.* London: Rich & Cowan, 1934.

Skinner, B. F. *Walden Two.* New York: Macmillan, 1970 [1948].

Smith, E. E. *The Skylark of Space.* London: Panther, 1974 [1928].

Smith, Michael G. *Rockets and Revolution: A Cultural History of Early Spaceflight.* Lincoln: University of Nebraska Press, 2014.

Smith, Roger. *The Fontana/Norton History of the Human Sciences.* London: Fontana/ New York: Norton, 1997.

Smith, Terry. *Making the Modern: Industry, Art, and Design in America.* University of Chicago Press, 1993.

Snow, C. P. *Science and Government.* Cambridge, MA: Harvard University Press, 1961.

The Two Cultures and a Second Look. Cambridge University Press, 1969.

Soddy, Frederick, *The Interpretation of Radium.* London: John Murray, 1909.

Sommerfeld, Vernon. *Speed, Space and Time.* London: Nelson, 1935.

Spanner, Edward Frank. *The Broken Trident.* London: Williams & Norgate, 1926.

The Naviators. London: Williams & Norgate, 1926.

Armaments and the Non-Combatant: To the 'Front-Line Troops' of the Future. London: Williams & Norgate, 1927.

The Harbour of Death. London: Williams & Norgate, 1927.

This Airship Business. London: Williams & Norgate, 1927.

Gentlemen Prefer Aeroplanes! London: E. F. Spanner, 1928.

The Tragedy of the 'R 101'. London: E. F. Spanner, 1931, 2 vols.

Stapledon, Olaf. *Last and First Men*. Reprinted London: Millennium, 1999 [1930].

Star Maker. Reprinted London: Millenium, 1999 [1937].

Stewart, Oliver. *Aeolus: Or the Future of the Flying Machine*. London: Kegan Paul, 1927.

Stokes, Simpson. *Air Gods' Parade*. London: Arthur Barron, 1935.

Stover, Leon. *The Prophetic Soul: A Reading of H. G. Wells's* Things to Come *together with his Film Treatment* Whither Mankind? *and the Postproduction Script*. Jefferson, NC: McFarland, 1987.

Science Fiction from Wells to Heinlein. Jefferson, NC: McFarland, 2002.

Sueter, Murray F. *Airmen or Noahs: Fair Play for the Airmen: The Great 'Neon' Air Myth Exposed*. London: Isaac Pitman & Sons, 1928.

Swann, Brenda and Francis Aprahamian (eds), *J. D. Bernal: A Life in Science and Politics*. London: Verso, 1999.

Swanwick, H. M. *Frankenstein and His Master: Aviation for World Service*. London: Women's International League, 1934.

Syon, Guillaume de. *Zeppelin! Germany and the Airships, 1900–1939*. Baltimore, MD: Johns Hopkins University Press, 2002.

Tatarsky, Daniel (ed.), *Eagle Annual: The 1950s*. London: Colin Frewin, 2007.

Taylor, Gordon Rattray. *The Biological Time Bomb*. London: Panther, 1969 [1968].

Teague, Walter Dorwin. *Design This Day: The Technique of Order in the Machine Age*. London: Studio Publations, 1947.

Thorold, Peter. *The Motoring Age: The Automobile and Britain, 1896–1939*. London: Profile Books, 2003.

Titterton, E. W. *Facing the Atomic Future*. London: Macmillan, 1956.

Tobey, Ronald C. *The American Ideology of National Science, 1919–1930*. University of Pittsburgh Press, 1971.

Toffler, Alvin. *Future Shock*. London: Bodley Head, 1970.

Trenn, Thaddeus. 'The Central Role of Energy in Soddy's Holistic and Critical Approach to Nuclear Science, Economics, and Social Responsibility.' *British Journal for the History of Science*, 12 (2009): 261–76.

Tunstall, Brian. *Eagles Restrained*. London: Allen & Unwin, 1936.

Turner, C. C. *Britain's Air Peril: The Danger of Neglect, together with Considerations on the Role of an Air Force*. London: Pitman, 1933.

Turner, Fred. *From Counterculture to Cyberculture: Stewart Brand, the Whole Earth Network, and the Rise of Digital Utopianism*. University of Chicago Press, 2006.

Turney, John. *Frankenstein's Footsteps: Science, Genetics and Popular Culture*. New Haven, CT: Yale University Press, 1998.

Van Pedroe-Savidge, E. *The Flying Submarine*. London: Arthur Stockwell, 1922.

Vogt, William. *Road to Survival*. London: Gollancz, 1949 [1948].

Voronoff, Serge. *The Conquest of Life*. London: Brentano's Ltd, 1928.

Wall, T. F. 'Seeking to Disrupt the Atom: Immeasurable Energy.' *Illustrated London News* (11 October 1924): 678.

Wallace, Graham. *Claude Grahame-White: A Biography*. London: Putnam, 1960.

Warren, Bill. *Keep Watching the Skies! American Science Fiction Movies of the Fifties.* Jefferson, NC: McFarland, 1982–86 (2 vols).

Warrick, Patricia S. 'Ethical Evolving Artificial Intelligence: Asimov's Computers and Robots' in Joseph D. Olander and Martin Harry Greenberg (eds), *Isaac Asimov.* Edinburgh: Paul Harris, 1977, pp. 174–200.

Wates, G. F. *All for the Golden Age: Or the Way of Progress.* London: Allen & Unwin, 1928.

Watkins, Tony. 'Piloting the Nation: Dan Dare in the 1950s' in Dudley Jones and Tony Watkins (eds), *A Necessary Fantasy? The Heroic Figure in Children's Popular Fiction.* New York: Garland Publishing, 2000, pp. 153–76.

Watson, John B. 'After the Family – What?' in V. F. Calverton and Samuel D. Schmalhausen (eds), *The New Generation: The Intimate Problems of Modern Parents and Children.* London: Allen & Unwin, 1930, pp. 55–73.

Watt, Donald (ed.), *Aldous Huxley: The Critical Heritage.* London: Routledge, 1975.

Weightman, Gavin. *Signor Marconi's Magic Box.* London: Harper Collins, 2003.

Weart, Spencer R. *Nuclear Fear: A History.* Cambridge, MA: Harvard University Press, 1988.

The Discovery of Global Warming. Cambridge, MA: Harvard University Press, 2003.

The Rise of Nuclear Fear. Cambridge, MA: Harvard University Press, 2012.

Weindling, Paul. *Health, Race and German Politics between National Unification and Nazism, 1870–1945.* Cambridge University Press, 1989.

Wells, Herbert George. *The Discovery of the Future: A Lecture Delivered at the Royal Institution on January 24, 1902.* London: T. Fisher Unwin, 1902.

The War in the Air. London: G. Bell, 1908.

Anticipations: Of the Reaction of Mechanical and Scientific Progress upon Human Life and Thought. New edn. With author's specially written introduction. London: Chapman & Hall, 1914.

The World Set Free: A Story of Mankind. London: Macmillan, 1914.

A Modern Utopia. London: Nelson, 1917. Reprint ed. Krishna Kuman, London: Deny, 1997.

The Outline of History: Being a Plain History of Life and Mankind. London: Newnes, 1920, 2 vols. Definitive edn., London: Cassell, 1923.

The Discovery of the Future. Revised edn. London: Jonathan Cape, 1925.

The Short Stories of H. G. Wells. London: Benn, 1926.

The Work, Wealth and Happiness of Mankind. London: Heinemann, 1932.

The Shape of Things to Come: The Ultimate Revolution. London: Hutchinson, 1933.

An Experiment in Autobiography: Discoveries and Conclusions of a Very Ordinary Brain (Since 1866). Reprinted London: Faber, 1984, 2 vols. [1934].

The Sleeper Awakes. Ed. Patrick Parrinder. London: Penguin Classics, 2005.

Things to Come: A Critical Text of the 1935 London First Edition. Ed. Leon Stover. Jefferson, NC: McFarland, 2007.

Wells, H. G., Julian S. Huxley and G. P. Wells, *The Science of Life.* Popular edn. London: Cassell, 1938 [1931].

Wendt, Gerald. *Science for the World of Tomorrow.* New York: Norton, 1939.

Westerman, Percy F. *The Airship 'Golden Hind'.* London: Partridge, n.d.

The Flying Submarine. London: James Nisbett, 1912.

Westfahl, Gary. *The Mechanics of Wonder: A History of the Idea of Science Fiction*. Liverpool University Press, 1998.

Cosmic Engineers: A Study of Hard Science Fiction. Westport, CT: Greenwood Press, 1996.

Westfahl, Gary, Wong Kin Yuen and Amy Kit-Sze Chan (eds), *Science Fiction and the Prediction of the Future*. Jefferson, NC: McFarland, 2011.

Whitworth, Michael H. *Einstein's Wake: Relativity, Metaphor, and Modernist Literature*. Oxford University Press, 2001.

Whyte, L. L. *Archimedes: Or the Future of Physics*. London: Kegan Paul, 1928.

Wilkes, M. V. 'A World Dominated by Computers' in Nigel Calder (ed.), *The World in 1984*. London: Penguin, 1964, vol. 1, pp. 147–50.

Williams, Beryl and Samuel Epstein. *The Rocket Pioneers*. London: Lutterworth Press, 1957.

Williams-Ellis, Clough (ed.), *Britain and the Beast*. London: Dent, 1937.

Wilson, Daniel H. *Where's My Jetpack? A Guide to the Amazing Science Fiction Future That Never Arrived*. New York: Bloomsbury Publishing, 2007.

Wilson, David. *Rutherford; Simple Genius*. London: Hodder & Stoughton, 1983.

Wilson, R. M. *Pygmalion: Or the Doctor of the Future*. London: Kegan Paul, 1926.

Winter, Frank H. *Prelude to the Space Age: The Rocket Societies, 1924–1940*. Washington: Smithsonian Institution Press, 1983.

Witkowski, J. A. 'Julian Huxley in the Laboratory: Embracing Inquisitiveness and Widespread Curiosity' in C. Kenneth Waters and Albert Van Helden (eds), *Julian Huxley: Biologist and Statesman of Science*. Houston, TX: Rice University Press, 1992, pp. 79–103.

Wittner, Lawrence S. *Confronting the Bomb: A Short History of the World Nuclear Disarmament Movement*. Stanford University Press, 2009.

Wohl, Robert. *A Passion for Wings: Aviation and the Western Imagination, 1908–1918*. New Haven, CT: Yale University Press, 1994.

The Spectacle of Flight: Aviation and the Western Imagination, 1920–1950. New Haven, CT: Yale University Press, 2005.

Worster, Donald. *Dust Bowl: The Southern Plains in the 1930s*. New York: Oxford University Press, 1979.

Worvill, Roy. *Exploring Space*. Loughborough: Wills & Hepworth, 1964.

Wright, H. W. S. *The Conquest of Cancer*. London: Kegan Paul, 1925.

Wright, Sydney Fowler. *The New Gods Lead*. London: Jarrolds, 1932.

Prelude in Prague: A Story of the War of 1938. London: George Newnes, 1935.

Wyndham, John. *The Chrysalids*. Reprinted Harmondsworth: Penguin, 1970 [1955].

Wyndham, John and Lucas Parkes. *The Outward Urge*. Harmondsworth: Penguin, 1962 [1959].

Wynne, Brian. *Rationality and Ritual: The Winscale Inquiry and Nuclear Decisions in Britain*. Chalfont St Giles, Bucks: British Society for the History of Science, 1982.

Zamyatin, Yevgeny. *We*. Trans. and introd. Clarence Browne. London: Penguin, 1993.

Index

264

Printed in the United States
By Bookmasters